Introductory Oceanography

Introductory Oceanography

Theobald Lane

R CALLISTO REFERENCE

www.callistoreference.com

Callisto Reference,
118-35 Queens Blvd., Suite 400,
Forest Hills, NY 11375, USA

Visit us on the World Wide Web at:
www.callistoreference.com

ISBN: 978-1-64116-583-9 (Hardback)

Cataloging-in-Publication Data

Introductory oceanography / Theobald Lane.
    p. cm.
Includes bibliographical references and index.
ISBN 978-1-64116-583-9
1. Oceanography. 2. Ocean. 3. Marine sciences. I. Lane, Theobald.
GC11.2 .I58 2022
551.46--dc23

Table of Contents

Preface

The study of the biological and physical aspects of the ocean is known as oceanography. It is a sub-discipline of Earth science. There are various aspects, which are studied within oceanography such as geophysical fluid dynamics, plate tectonics, ocean currents and ecosystem dynamics. There are primarily four sub-disciplines within oceanography, namely, chemical oceanography, biological oceanography, physical oceanography and geological oceanography. Chemical oceanography further involves ocean acidification where the pH level of the ocean is studied. Some of the numerous fields where oceanography is applied are geography, climatology, chemistry, biology, astronomy and hydrology. The topics included in this book on oceanography are of utmost significance and bound to provide incredible insights to readers. It presents researches and studies performed by experts across the globe. Coherent flow of topics, student-friendly language and extensive use of examples make this book an invaluable source of knowledge.

A foreword of all chapters of the book is provided below:

Chapter 1 - The scientific study of the oceans including its physical and chemical properties is known as oceanography. The study of the oceans helps in a better understanding of the weather changes. This chapter delves into the significant aspects of the ocean and oceanography; **Chapter 2 -** Oceanography can be further divided into several branches. These include physical oceanography, biological oceanography, chemical oceanography, paleoceanography and geologic oceanography. All these branches have been explored in detail in this chapter; **Chapter 3 -** The ocean ecosystem is a vast area of study. It can broadly be classified into open ocean ecosystem, deep ocean ecosystem, coral reef ecosystem, shoreline ecosystems, etc. These diverse types of ocean ecosystems have been carefully analyzed in this chapter; **Chapter 4 -** There are numerous phenomena which occur within the ocean ecosystem. Some of these are sea foam, tidal bores, red tide, upwelling, whirlpool, etc. These phenomena have been discussed in detail in the following chapter; **Chapter 5 -** Some of the diverse aspects of the ocean studied within oceanography are ocean current, tides and waves. It also studies ocean basins and sediments. The topics elaborated in this chapter will help in gaining a better perspective about these diverse aspects of the ocean.

At the end, I would like to thank all the people associated with this book devoting their precious time and providing their valuable contributions to this book. I would also like to express my gratitude to my fellow colleagues who encouraged me throughout the process.

Theobald Lane

Chapter 1

Oceanography: An Introduction

The scientific study of the oceans including its physical and chemical properties is known as oceanography. The study of the oceans helps in a better understanding of the weather changes. This chapter delves into the significant aspects of the ocean and oceanography.

Ocean

An ocean is a major body of saline water, and a principal component of the hydrosphere. Approximately 70 percent of the Earth's surface (an area of some 361 million square kilometers (139 million square miles) is covered by saline water forming one continuous body that is customarily divided into several principal oceans and smaller seas. More than half of this area is over 3,000 meters (9,800 ft) deep. Average oceanic salinity is around 35 parts per thousand (ppt) (3.5 percent), and nearly all seawater has a salinity in the range of 31 to 38 parts per thousand with salinity varying according to such factors as precipitation, evaporation, melting of sea ice, and river inflow.

The world ocean, an integral part of global climate, is constantly changing, absorbing heat from the sun and cooling through evaporation, dissolving and releasing carbon dioxide, and moving in great conveyor belt currents transferring heat and moisture toward the poles from the tropics and deep below the surface returning cold water to the tropics. From ocean breezes to monsoons, hurricanes, summer rains, and winter fog, the oceans' heat and water vapor are constantly affecting life on land, even far from the ocean shore.

The plants and animals living in the world ocean provide human beings with a vast food resource that has tragically been threatened by overexploitation and pollution caused by human activity. Establishing proper use of the ocean will require international cooperation and coordination aligned with the values of co-existence with nature and mutual prosperity for all humankind. As the place where national sovereignties interface with internationally sovereign waters, and where many aquatic species freely traverse the boundaries between the two, the world ocean is a critically important arena in which to resolve issues that have heretofore hindered progress toward a global peace.

The World Ocean is one global, interconnected body of salt water comprising the world's five oceans – Atlantic, Pacific, Indian, Arctic, and Southern oceans. The concept of a global ocean as a continuous body of water with relatively free interchange among its parts is of fundamental importance to oceanography.

Major oceanic divisions are defined by various criteria, including the shores of continents and various archipelagos. These divisions are (in descending order of size) the Pacific Ocean, the Atlantic Ocean, the Indian Ocean, the Southern Ocean (which is sometimes subsumed as the southern portions of the Pacific, Atlantic, and Indian Oceans), and the Arctic Ocean (which is sometimes considered a sea of the Atlantic). The Pacific and Atlantic may be further subdivided by the equator into northerly and southerly portions.

Smaller regions of the oceans are called seas, gulfs, bays, and so forth. In addition, there are some smaller bodies of saltwater that are totally landlocked and not interconnected with the World Ocean, such as the Caspian Sea, the Aral Sea, and the Great Salt Lake. Although some of them are referred to as "seas," they are actually salt lakes.

Geological Perspective

Geologically, an ocean is an area of oceanic crust covered by water. Oceanic crust is the thin layer of solidified volcanic basalt that covers the Earth's mantle where there are no continents. From this perspective, there are three oceans today: the World Ocean and two seas, the Caspian and the Black Sea, the latter two of which were formed by the collision of the Cimmerian plate with Laurasia. The Mediterranean Sea is very nearly a discrete ocean, being connected to the World Ocean only through the eight-mile-wide Strait of Gibraltar, which several times over the last few million years has been closed off completely due to tectonic movement of the African continent. The Black Sea is connected to the Mediterranean through the Bosporus, but this is in effect a natural canal cut through continental rock some 7,000 years ago, rather than a piece of oceanic sea floor like that underlying the Strait of Gibraltar.

Physical Properties

The area of the World Ocean is approximately 361 million square kilometers (139 million sq mi); its volume is approximately 1,300 million cubic kilometers (310 million cu mi); and its average depth is 3,790 meters (12,430 ft). Nearly half of the world's marine waters are over 3,000 meters (9,800 ft) deep. The vast expanses of Deep Ocean (depths over 200 m) cover more than half of the Earth's surface.

The total mass of the hydrosphere is about 1.4×10^{21} kilograms, which is about 0.023 percent of the Earth's total mass. Less than 2 percent is freshwater, the rest is saltwater, mostly in the ocean.

Color

A common misconception is that the oceans are blue primarily because the sky is blue. In fact, water has a very slight blue color that can only be seen in large volumes. Although the sky's reflection does contribute to the blue appearance of the surface, it is not the primary cause. The primary cause is the absorption of red photons from the incoming light by the nuclei of water molecules. The absorption by the nuclei is an anomaly because it occurs through a vibrational change, whereas all other known examples of color in nature result from electronic dynamics.

Regions

Oceans are divided into numerous regions depending on physical and biological conditions: The pelagic zone, which includes all open ocean regions, is often subdivided into further regions categorized by depth and abundance of light. The photic zone covers the oceans from surface level to 200 meters down. This is the region where photosynthesis occurs most commonly and therefore where the largest biodiversity in the ocean lives. Since plants can only survive through photosynthesis, any life found lower than this must either rely on organic detritus floating down from above (marine snow) or find another primary source such as hydrothermal vents in what is known as the aphotic zone (all depths exceeding 200 m). The pelagic part of the photic zone is known as the epipelagic. The pelagic part of the aphotic zone can be further divided into regions that succeed each other vertically. The mesopelagic is the uppermost region, with its lowermost boundary at a thermocline of 10°C, which, in the tropics generally lies between 700 meters and 1,000 meters (2,297 and 3,280 feet). Directly below that is the bathypelagic lying between 10°C and 4°C, or between 700 or 1,000 meters (2,297 and 3,280 feet) and 2,000 or 4,000 meters (6,560 or 13,123 feet). Lying along the top of the abyssal plain is the abyssal pelagic, whose lower boundary lies at about 6,000 meters (19,685 feet). The final zone falls into the oceanic trenches, and is known as the hadalpelagic. This lies between 6,000 meters and 10,000 meters (19,685 and 32,808 feet) and is the deepest oceanic zone.

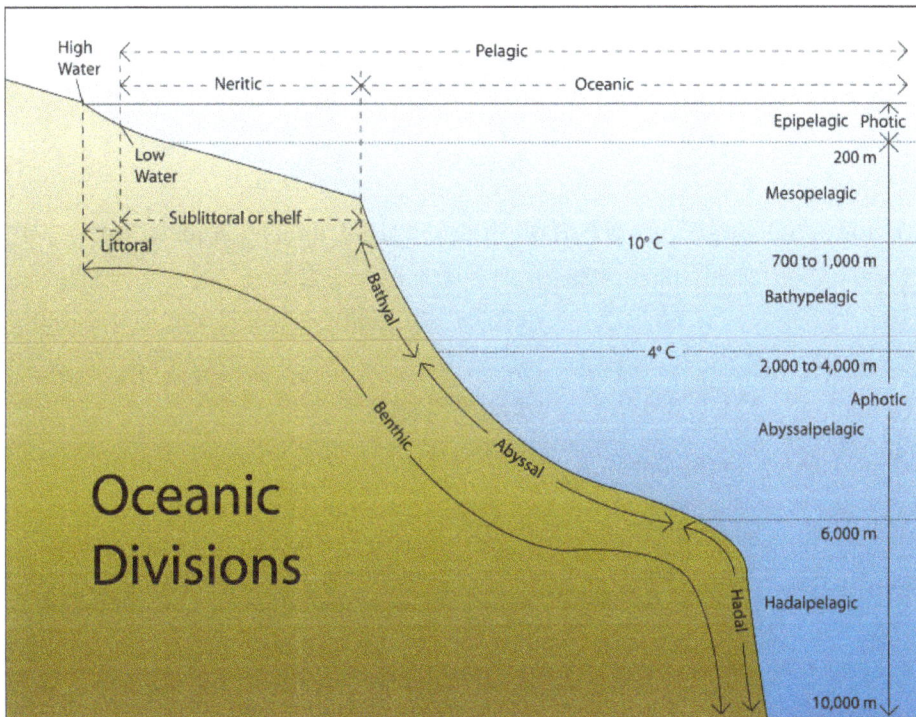

Oceanic subdivisions.

Along with pelagic aphotic zones there are also seafloor or benthic aphotic zones corresponding to the three deepest zones: The bathyal zone covers the continental slope and the sides of the mid-ocean ridge down to about 4,000m. The abyssal zone covers the abyssal plains between 4,000 and 6,000m. Lastly, the hadal zone corresponds to the hadalpelagic zone which is found in the oceanic trenches.

The pelagic zone can also be split into two subregions, the neritic zone and the oceanic zone. The neritic encompasses the water mass directly above the continental shelves, while the oceanic zone includes all the completely open water. In contrast, the littoral zone covers the region between low and high tide and represents the transitional area between marine and terrestrial conditions. It is also known as the intertidal zone because it is the area where tide level affects the conditions of the region.

Climate

One of the most dramatic forms of weather occurs over the oceans: Tropical cyclones (also called hurricanes, typhoons, tropical storms, cyclonic storms, and tropical depressions depending upon where the system forms). A tropical cyclone feeds on the heat released when moist air rises and the water vapor it contains condenses. Tropical cyclones can produce extremely powerful winds and torrential rain, high waves and damaging storm surge. Although their effects on human populations can be devastating, tropical cyclones also relieve drought conditions. They also carry heat and energy away from the tropics and transport it towards temperate latitudes, which makes them an important part of the global atmospheric circulation mechanism. Tropical cyclones help to maintain equilibrium in the Earth's troposphere, and to maintain a relatively stable and warm temperature worldwide.

Ocean currents greatly affect Earth's climate by transferring warm or cold air and precipitation to coastal regions, where they may be carried inland by winds. The Antarctic Circumpolar Current encircles that continent, influencing the area's climate and connecting currents in several oceans.

Ecology

The oceans are home to a large number of plant and animal species, including:

- Radiata,
- Fish,
- Cetacea such as whales, dolphins and porpoises,
- Cephalopods such as the octopus,
- Crustaceans such as lobsters and shrimp,
- Marine worms,
- Plankton,
- Krill.

Endangered Species

Until recently, the ocean appeared to be a vast and infinite source of food, invulnerable to exploitation. In contrast, the reality is that the populations of many species living in the ocean are decreasing rapidly. NOAA has jurisdiction over 157 endangered and threatened marine species, including 63 foreign species. Marine life is vulnerable to problems such as overexploitation, pollution, habitat destruction, and climatic changes. Air-breathing animals such as whales, turtles and manatees are often caught in fishing nets or injured by boats. Species such as birds and turtles that lay their eggs on land lose their nurseries to coastal development, and the spawning grounds of fish are eliminated by alterations to inland waterways such as dams and diversion canals. Pollution from ships, raw sewage, and ground run-off create nutrient overloads in the waters or poison corals and the small organisms that feed larger animals.

Economy

The oceans are essential to transportation: most of the world's goods are moved by ship between the world's seaports. The Panama and Suez canals allow ships to pass directly from one ocean into another without having to circumnavigate South America and Africa respectively.

The oceans are an important source of valuable foodstuffs through the fishing industry. Aquaculture, an expanding industry, achieves increased production of specific species under controlled conditions while also relying heavily on the oceans as a source of feed stock for the farmed fish.

During the twentieth century, exploitation of natural resources under the sea began with the drilling of oil wells in the sea bed. During the 1950s, companies began to research the possibility of mining the ocean floor for mineral resources such as diamonds, gold, silver, manganese nodules, gas hydrates and underwater gravel. In 2005, Neptune Resources NL, a mineral exploration company, applied for and was granted 35,000 km² of exploration rights over the Kermadec Arc in New Zealand's Exclusive Economic Zone to explore for seafloor massive sulfide deposits, a potential new source of lead-zinc-copper sulfides formed from modern hydrothermal vent fields.

The oceans are also a vital resource for tourism. In every country that has a coastal boundary beaches are favorite places for relaxation, water sports, and leisure. With the advent of train and air travel, millions of tourists began to visit beaches in countries with warm climates. Many developing nations rely on tourism to their beach resorts as a major element of their economies. Travel on large luxury cruise ships is becoming increasingly popular.

Ancient Oceans

Continental drift has reconfigured the Earth's oceans, joining and splitting ancient oceans to form the current ones. Ancient oceans include:

- Bridge River Ocean, the ocean between the ancient Insular Islands and North America.
- Iapetus Ocean, the southern hemisphere ocean between Baltica and Avalonia.
- Panthalassa, the vast world ocean that surrounded the Pangaea supercontinent.
- Rheic Ocean.
- Slide Mountain Ocean, the ocean between the ancient Intermontane Islands and North America.
- Tethys Ocean, the ocean between the ancient continents of Gondwana and Laurasia.
- Khanty Ocean, the ocean between Baltica and Siberia.
- Mirovia, the ocean that surrounded the Rodinia supercontinent.
- Paleo-Tethys Ocean, the ocean between Gondwana and the Hunic terranes.
- Proto-Tethys Ocean.

- Pan-African Ocean, the ocean that surrounded the Pannotia supercontinent.

- Superocean, the ocean that surrounds a global supercontinent.

- Ural Ocean, the ocean between Siberia and Baltica.

Extraterrestrial Oceans

Earth is the only known planet with liquid water on its surface and is certainly the only one in our Solar System. Astronomers think, however, that liquid water is present beneath the surface of the Galilean moons Europa, and (with less certainty) Callisto and Ganymede. Geysers have been observed on Enceladus, though they may not involve bodies of liquid water. Other icy moons such as Triton may have once had internal oceans that have now frozen. The planets Uranus and Neptune may also possess large oceans of liquid water under their thick atmospheres, though their internal structure is not well understood at this time.

There is currently much debate over whether Mars once had an ocean of water in its northern hemisphere, and over what happened to it if it did. Recent findings by the Mars Exploration Rover mission indicate: Mars probably had some long-term standing water in at least one location, but its extent is not known. Astronomers believe that Venus had liquid water and perhaps oceans in its very early history. If they existed, all traces of them seem to have vanished in later resurfacing of the planet.

Liquid hydrocarbons are thought to be present on the surface of Titan, though it may be more accurate to describe them as "lakes" rather than an "ocean." The Cassini-Huygens space mission initially discovered only what appeared to be dry lakebeds and empty river channels, suggesting that Titan had lost what surface liquids it might have had. Cassini's more recent fly-by of Titan has yielded radar images strongly suggestive of hydrocarbon lakes near the Polar Regions where it is colder. Scientists also think it likely that Titan has a subterranean water ocean under the mix of ice and hydrocarbons that forms its outer crust.

Gliese 581 c, one of the extrasolar planets that have been found in recent years, is at the right distance from its sun for liquid water to exist on the planet's surface. Since the alignment of Gliese 581 c's orbit in relation to the viewing angle from earth precludes a visible transit by the planet of its sun, there is no way to know if the planet does have liquid water. Some researchers have suggested that the extrasolar planet HD 209458b may have water vapor in its atmosphere, but this view is currently being disputed. The extrasolar planet Gliese 436 b is believed to have 'hot ice', i.e., ice existing under conditions of greater gravity than on earth and hence with a higher melting temperature than on earth. If water molecules exist on either HD 209458b or Gliese 436 b, they are likely to be found also on other planets at a suitable temperature, meaning that there would be some further reason to hope someday to find another planet besides Earth with water ocean.

Open Ocean Deep Sea

The deep sea comprises the seafloor, water column and biota therein below a specified depth contour. There are differences in views among experts and agencies regarding the appropriate depth to delineate the "deep sea".

Benthic Realm

Deep-sea Margins

The global continental margins extend for ~150,000 km and encompass estuarine, open coast, shelf, canyon, slope, and enclosed-sea ecosystems. Deep-sea margins are those areas that lie beyond the shelf break, where the seafloor slopes down to the continental rise at abyssal depths, and encompasses bathyal depths. Numerous canyons and channels incise the continental slope, often featuring cold-water coral reefs or oxygen minimum zones (OMZs) as distinct habitats along the deep margin. Sediment covers much of the deep continental margin, but with exposed bedrock in areas where topography is too steep for sediment accumulation (e.g., steep canyon walls) or where sediment is washed away (e.g., parts of seamounts). Different faunas inhabit soft- and hard-bottom substrates.

Relative to their area, the margins account for a disproportionately large fraction of global primary production (10-15 per cent), nutrient recycling, carbon burial (>60 per cent of total settling organic carbon), and fisheries production. They are also exceptionally dynamic systems with ecosystem structures that can oscillate slowly or shift abruptly, but rarely remain static.

Status and Trends for Biodiversity

In the well-studied North Atlantic, local macrofaunal (300 μm-3 cm) species diversity on the continental slope exceeds that of the adjacent continental shelf, and estimates of bathyal diversity in other parts of the world ocean are comparably high, but local environmental conditions drive regional differences: e.g., the Gulf of Mexico, the Norwegian and Mediterranean Seas, the Eastern. Pacific and the Arabian Sea. Most researchers agree that habitat heterogeneity on different spatial scales drives high diversity along the margins and that margins often exhibit upwelling and increased production that enhances biodiversity. Nonetheless, excess food availability can reduce diversity.

Depth-related species diversity gradients in macrofauna often peak unimodally at mid-bathyal depths of about 1500-2000 m, although shallower peaks in diversity have been observed in Arctic waters for bivalves, polychaetes, gastropods and cumaceans, as well as for the entire macrofauna and some meiofauna (32 μm-1000 μm). Even regions with very low diversity can host highly specialized species (e.g., OMZs) and contribute to overall margin diversity Thus, throughout their depth gradient, continental margin slope areas exhibit the highest macrofaunal diversity and offer a potentially important refuge against future climate change, as mobile organisms could migrate upslope or downslope in search of suitable conditions.

The diversity of meiofauna (32 μm-1,000 μm) exceeds that of the macrofauna and their diversity generally increases with depth; however, groups such as foraminifera and ostracods exhibit unimodal peaks in diversity. Meiofaunal diversity may decline or increase with increasing bathyal depths, generally driven by food availability and intensity and regularity of disturbance regimes, as well as by temperature and local environmental conditions.

Russian and Scandinavian deep-sea expeditions described peak benthic megafaunal (>3 cm) diversity at mid-bathyal depths as early as the 1950s and 1960s, despite observing much lower megafaunal than meio- and macrofaunal diversity. Sponges, cnidarians, crustaceans (decapods and isopods) and echinoderms (echinoids, asteroids, crinoids, holothurians) all display this pattern, however later studies confirmed the pattern for some megafaunal invertebrates, but showed a decline or even increase in diversity with increasing depth for some taxa. Evidence to date suggests lower species richness in deep-sea bacterial communities than in coastal benthic environments, with the caveat that deep-sea environments remain underexplored. However, the presence of extreme environments in the deep sea which have high phylogenetic diversity promises a rich source of bacterial diversity and genetic innovation.

Several faunal groups also exhibit latitudinal gradients in species diversity: diversity of crustaceans, molluscs and foraminifera declines poleward, whilst others such as nematodes respond to phytodetrital input. Latitudinal gradients have also been identified in bacteria but recent modelling indicates peak bacterial richness in temperate areas in winter. The effect of seasons on macro-ecological patterns in the microbial ocean warrants continued investigation to test the mechanisms that underlie latitudinal patterns in different fauna.

Broad-scale depth and latitudinal patterns in benthic diversity are modified regionally by a variety of environmental factors operating at different scales. For example, OMZs strongly affect diversity where they impinge on the seafloor. OMZs typically occur between 200 m and 1000 m, often at major carbon burial sites along the continental margins where high productivity results in high carbon fluxes to the seafloor and low oxygen. The organic-rich sediments of these regions often support mats of large sulphide-oxidizing bacteria (Thioploca, Beggiatoa, Thiomargarita), and high-density, low-diversity metazoan assemblages. Protists are also well represented in OMZs such as the Cariaco Basin, where representatives of all major protistan clades occur. Depressed diversity near OMZs centres favours taxa that can tolerate hypoxia, such as nematodes and certain annelids and foraminifera. Other taxa that cannot tolerate low-oxygen conditions may aggregate at the OMZs fringes where food is often abundant.

Major Pressures

Multiple anthropogenic influences affect deep-sea habitats located close to land (e.g., canyons, fjords, upper slopes when continental shelves are very narrow), including organic matter loading, mine tailings disposal, litter, bottom trawling and overfishing, enhanced or decreased terrestrial input, oil and gas exploitation and, potentially in future, deep-sea mining. Fishing on margins can also have indirect ecological effects at deeper depths. These anthropogenic influences can modify deep-margin habitats through physical smothering and disturbance, sediment resuspension, organic loading, and toxic contamination and plume formation, with concomitant losses in biodiversity, declining energy flow back to higher trophic levels, and impacts on physiology from exposure to toxic compounds (e.g., hydrocarbons, polycyclic aromatic hydrocarbons (PAHs), heavy metals).

Abyss

Status and Trends for Biodiversity

The abyss (~3-6 km water depth) encompasses the largest area on Earth. Its vast areas of seafloor plains and rolling hills are generally covered in fine sediments with hard substrates associated with manganese nodules, rock outcrops and topographic highs (e.g. seamounts). The absence of in situ primary production in this comparatively stable habitat characterize an ecosystem adapted to a limiting and variable rain of particulate detrital material that sinks from euphotic zones. Nonetheless, the abyss supports higher levels of alpha and beta diversity of meiofauna, macrofauna and megafauna than was recognized only decades ago. The prevalence of environmental DNA preserved in the deep sea biases estimates of richness, at least in the microbial domain, adding a challenge to biodiversity study in the abyss using molecular methods.

Despite poorly known biodiversity patterns at regional to global scales (especially regarding species ranges and connectivity), some regions, such as the abyssal Southern Ocean and the Pacific equatorial abyss, are likely to represent major reservoirs of biodiversity.

Major Pressures

The food-limited nature of abyssal ecosystems, and reliance on particulate organic carbon (POC) flux from above, suggest that all groups, from microbes to megafauna, will be highly sensitive to changes in phytoplankton productivity and community structure, and especially to changes in the quantity and quality of the export flux. Climate warming in some broad areas may increase ocean stratification, reduce primary production, and shift the dominant phytoplankton community structure from diatoms to picoplankton, and reduce export efficiency, driving biotic changes over major regions of the abyss, such as the equatorial Pacific. However the effects of climate change, including ocean warming, on biodiversity are likely to vary regionally and among species groups in ways that are poorly resolved with current models and knowledge of ecosystem dynamics in the deep sea.

The Hadal Zone

The Hadal zone, comprising ocean floor deeper than 6000 m, encompasses 3,437,930 km2, or less than 1 per cent of total ocean area and represents 45 per cent of its depth and related gradients. Over 80 separate basins or depressions in the sea floor comprise the hadal zone, dominated by 7 great trenches (>6500 m) around the margins of the Pacific Ocean, five of which extend to over 10 km depth: the Japan-Kuril-Kamchatka, Kermadec, Tonga, Mariana, and Philippine trenches. The Arctic Ocean and Mediterranean Sea lack hadal depths. These trenches are often at the intersection of tectonic plates, exposing them as potential epicentres of severe earthquakes which can directly cause local and catastrophic disturbance to the trench fauna.

Status and Trends for Biodiversity

Although the hadal zone contains a wide range of macro- and megafaunal taxa (cnidarians, polychaetes, bivalves, gastropods, amphipods, decapods, echiurids, holothurians, asteroids, echinoids, sipunculids, ophiuroids and fishes, all trenches occur below the Carbonate Compensation Depth (CCD), reducing the numbers of calcified protozoan and metazoan species found there. Chemosynthetic seep biota, including vesicomyid and thyasirid clams, occur in hadal depths in the Japan Trench; the deepest known methane seeps and associated communities are found at 7,434 m in this area. Cold seep communities also commonly occur in trench areas, such as the Aleutian and Kuril Trenches. Benthic foraminifera are among the most widespread taxa at hadal depths and include calcareous, large agglutinated, and organic walled species. Abundant metazoan meiofaunal taxa, such as nematodes, at hadal depths may exceed those found at bathyal depths by 10-fold;

small numbers of ostracods, halacarids, cumaceans, kinorhynchs, and meiofaunal-sized bivalves are also found there. Nematode and copepod communities in trenches differ greatly from those found at bathyal and abyssal depths, driven by opportunistic taxa and meiofaunal dwarfism in trench systems.

Although not yet well quantified, and the mechanisms remain to be discerned, higher densities of fauna and respiration have been found at trench axis points than would be expected from a purely vertical rain of POC flux.The exact number of species in trenches is not known, but the few quantitative studies made so far suggest that diversity is lower compared to diversity at abyssal depths. Reasons for the lower diversity levels are not well understood but the high pressure, relatively high food supply and organic matter accumulation, relatively elevated temperature (due to adiabatic heating), or a combination thereof may attenuate trench diversity.

Sampling to date suggests that hadal basins are populated by a higher proportion of endemic species compared to much shallower waters, species that can survive the extreme hydrostatic pressure and, in some instances, remoteness from surface food supply. Physiological and other evidence suggests that fishes cannot survive at depths greater than 8000 m; the deepest hadal fish, the liparids (snail-fish), are unique to each trench system. Decapod crustaceans have been observed only to 8200 m (Gallo in revision).

At depths over 8000 m, scavenging amphipod crustaceans dominate the mobile megafauna, along with potential predators, including penaeid shrimp, princaxelid amphipods and ulmarid jellyfish, as observed in the New Britain Trench and the Sirena Deep (Mariana Trench). Comparison of scavenging and epibenthic/demersal biota suggests that density, diversity, and incidence of demersal (near bottom) lifestyles all increase with greater food supply.

Wide separation between trenches in the northern and southern hemispheres and between the different oceans has likely facilitated speciation to result in distinct assemblages of fauna in each hadal basin. Some 75 per cent of the species in Pacific Ocean trenches may be endemic to each trench. Despite their remoteness from the surface, many hadal trenches are close to land and receive organic inputs from terrestrial and coastal sources, yielding higher mega-, macroand meio-faunal densities than expected for greater depths.

Major Pressures

The proximity of some trenches to land also increases their vulnerability to human activity in terms of dumping of materials and effluents, as well as from disaster debris, run off from land and pollution from ships. Some of these items, including anthropogenic litter, have been observed down to 7,200 m depth. Evidence for the vulnerability of trench fauna is also provided by the levels of the radioisotope 134Cs detected in sediments in the Japan Trench, four months after the Fukushima Dai-ichi nuclear disaster.

Knowledge Gaps

Trenches are arguably the most difficult deep-sea environments to access and current facilities are very limited worldwide, and consequently knowledge of their biodiversity is particularly incomplete. In general, biodiversity patterns of non-nematode meiofauna and non-foraminiferal protists are especially poorly known in the deep sea.

Most information about biodiversity in the deep sea is for the predominant softsubstrate habitats. However, hard substrates abound in the deep sea in nearly all settings, and organisms that cannot be seen in a photograph or video image are hard to sample and study quantitatively. Thus knowledge of small-taxon biodiversity is best developed for deep-sea sediments.

Beyond cataloguing diversity, even in those systems we have characterized, almost nothing is known about the ranges of species, connectivity patterns or resilience of assemblages and their sensitivity to climate stressors or direct human disturbance. There is also currently a lack of appropriate tools to adequately evaluate human benefits that are derived from the deep sea.

Pelagic Realm

Status and Trends for Biodiversity

Between the deep-sea bottom and the sunlit surface waters are the open waters of the deep pelagic or "midwater" environment. This huge volume of water is the least explored environment on our planet. The deep pelagic realm is very diffuse, with generally low apparent abundances of inhabitants, although recent observations from submersibles indicate that some species may concentrate into narrow depth bands.

The major physical characteristics structuring the pelagic ecosystems are depth and pressure, temperature, and the penetration of sunlight. Below the surface zone (or epipelagic, down to about 200 m), the deep layer where sunlight penetrates with insufficient intensity to support primary production, is called the mesopelagic zone. In some geographic areas, microbial degradation of organic matter sinking from the surface zone results in low oxygen concentrations in the mesopelagic, called OMZs. This mesopelagic zone is a particularly important habitat for fauna controlling the depth of CO_2 sequestration.

Below the depth to which sunlight can penetrate (about 1,000 m) is the largest layer of the deep pelagic realm and by far the largest ecosystem on our planet, the bathypelagic region. This comprises almost 75 per cent of the volume of the ocean and is mostly remote from the influence of the bottom and its communities. Temperatures there are usually just a few degrees Celsius above zero. The boundary layer where both physical and biological interactions with the bottom occur is called 'benthopelagic'.

The transitions between the various vertical layers are gradients, not fixed surfaces; hence ecological distinctions among the zones are somewhat blurred across the transitions. Recent surveys have shown a great deal of connectivity between the major pelagic depth zones. The abundance and biomass of organisms generally varies among these layers from a maximum near the surface, decreasing through the mesopelagic, to very low levels in the bathypelagic, increasing somewhat in the benthopelagic. Although abundances are low, because such a huge volume of the ocean is bathypelagic, even species that are rarely encountered may have very large total population numbers.

The life cycles of deep-sea animals often involve shifts in vertical distribution among developmental stages. Even more spectacular are the daily vertical migrations of many mesopelagic species. This vertical migration may increase physical mixing of the ocean water and also contributes to a "biological pump" that drives the movement of carbon compounds and nutrients from the surface waters into the deep ocean.

Sampling the deep pelagic biome shares the logistical difficulties of other deep-sea sampling, compounded by the extremely large volume and temporal variability of the environment and the widely dispersed populations of its inhabitants. New species continue to be discovered regularly. Whereas scientific information on the composition of mesopelagic assemblages is rapidly improving, very little is known of the structure of the deeper lower bathyal and abyssal pelagic zones.

Possibly because of high mobility and transport by ocean current, the overall diversity of species seems to be less than that found in other ecosystems. However, the number of distinct major evolutionary groups (i.e., phyla, classes, etc.) found in the deep pelagic is high.

Studies of microbes and their roles in the deep pelagic ecosystems are just beginning to reveal the great diversity of such organisms. The species richness of deep ocean bacteria surpasses that of the surface open ocean.

As is true in other pelagic systems, crustaceans make up a large percentage of the deep zooplankton in both abundance and numbers of species. These crustaceans include numerous and diverse copepods, amphipods, ostracods and other major groups. Some groups, like arrow worms, are almost all pelagic and are important in deep waters. Large gelatinous animals, including comb jellies, jellyfishes, colonial siphonophores, salps and pyrosomes, are extremely important in deep pelagic ecosystems.

The strong swimmers of the deep pelagic, the "nekton", include many species of fishes and some sharks, crustaceans (shrimps, krill, and other shrimplike animals), and cephalopods (including squids, "dumbo" and other octopods, and "vampire squids"). In terms of global fish abundance, deep pelagic fishes are by far the numerically dominant constituents; the genus Cyclothone alone outnumbers all coastal fishes combined and is likely to be the most abundant vertebrate on earth. Furthermore, at an estimated ~1,000 million tons, mesopelagic fishes dominate the world's total fish biomass and constitute a major component of the global carbon cycle. Acoustic surveys now suggest that an accurate figure of mesopelagic fish biomass may be an order of magnitude higher. When bathypelagic fish biomass is included, deep pelagic fish biomass is likely to be the overwhelming majority of fish biomass on Earth. The deep pelagic fauna is also important prey for mammals (toothed whales and elephant seals) and even birds (emperor penguins) and reptiles (leatherback sea turtles). The amount of deep-sea squids consumed by sperm whales alone annually has been estimated to exceed the total landings of fisheries worldwide.

Horizontal patterns exist in the global distribution of deep pelagic organisms. However, the faunal boundaries of deep pelagic assemblages are less distinct than those of near-surface or benthic assemblages. Generally, the low-latitude oligotrophic regimes that make up the majority of the global ocean house more species than higher-latitude regimes. Some major oceanic frontal boundaries, such as the polar and subpolar fronts, extend down into deep waters and appear to form biogeographic boundaries, although the distinctness of those boundaries may decrease with increasing depth.

The dark environment also means that production of light by bioluminescence is almost universal among deep pelagic organisms. Some animals produce the light independently, whereas others are symbiotic with luminescent bacteria.

Major Pressures

A fundamental biological characteristic throughout the deep pelagic biome is that little or no primary production occurs and deep pelagic organisms are dependent on food produced elsewhere. Therefore, changes in surface productivity will be reflected in changes in the deep midwater. When midwater animals migrate into the surface waters at night, they are subjected to predation by near-surface species. Shifts in the abundance of those predators will affect the populations of the migrators and, indirectly, the deeper species that interact with the vertical migrators at their deeper daytime depths. Either or both of these effects may be caused by global climate change, fishing pressure and the impact of pollutants in surface waters.

Climate change wills likely increase stratification caused by warming of surface waters and expanded OMZs resulting from the interaction of shifts in productivity with increased stratification. If the so-called conveyor-belt of global circulation weakens, transport of oxygen by the production of deep water will affect the entire deep sea. The biomass of mesopelagic fishes in the California Current, for instance, has declined dramatically during recent decades of reduced midwater oxygen concentrations. Furthermore, increases in carbon dioxide resulting in acidification may affect diverse deep pelagic animals, including pteropods (swimming snails) and crustaceans which use calcium carbonate to build their exoskeletons, fishes that need it for internal skeletons, and cephalopods for their balance organs. Acidification also changes how oxygen is transported in the blood of animals and those living in areas of low oxygen concentration may therefore be less capable of survival and reproduction.

Few fisheries currently target deep pelagic species, but fisheries do affect the ecosystem. Whaling reduced worldwide populations of sperm whales and pilot whales to a small fraction of historical levels. Similarly, fisheries for surface predators such as sharks, tunas and billfishes, and on seamounts, reduce predation pressure, particularly on vertical migrators like squids and lantern fishes.

Increasing extraction of deep-sea hydrocarbon resources increases the likelihood of accidental deep release of oil and methane, as well as the deep use of dispersants to minimize apparent effects of such spills at the surface.

Knowledge Gaps

Any summary of deep pelagic ecosystems emphasizes how little is known, especially relative to coastal systems. Sampling has been intensively conducted in only a few geographic areas, using selective methods, each of which illuminates only a fraction of the biodiversity. Sampling at lower bathyal or abyssal depths has been limited, and virtually nothing is known about pelagic fauna associated with deep trenches. There is also limited knowledge of the performance of conservation and management measures developed for coastal and shelf marine ecosystems when applied in deep ocean systems characterized by large spatial scales and variable but sometimes vertically and/or horizontally high-mobility organisms, and incomplete knowledge of ecosystem structure and processes.

Special Areas Typical for the Open Ocean Deep Sea

Ocean Ridges

The Mid-Ocean Ridge system is a continuous single feature on the earth's surface extending ca. 50,000 km around the planet; it defines the axis along which new oceanic crust is generated at tectonic plate boundaries. The ridge sea floor is elevated above the surrounding abyssal plains, reaching the sea surface at mid-ocean islands, such as Iceland, the Azores and Ascension Island in the Atlantic Ocean, Easter Island and Galapagos in the Pacific Ocean. Typically there is a central axial rift valley bounded by ridges on both sides. A series of sediment-covered terraces slope down on the two sides of the ridge axis to the abyssal plains. The global ridge system, including associated island slopes, seamounts and knolls, represents a vast area of mid-ocean habitat at bathyal depths, accessible to fauna normally associated with narrow strips of suitable habitat on the continental slopes. The ocean ridges sub-divide the major ocean basins, but fracture zones at intervals permit movement of deep water and abyssal organisms between the two sides of the ridge.

Much attention has been directed to the importance of Mid-Ocean Ridges as sites of the hydrothermal vents and their unique fauna found close to the geothermally active ridge axis. However, the total area of hydrothermal vents is small and the dominant fauna on the mid-ocean ridges is made up of typical bathyal species known from adjacent continental margins. The biomass of benthic fauna and demersal fishes on the ridges is generally similar to that found at corresponding depths on the nearest continental slopes. New species, potentially endemic to mid-ocean ridges, have been discovered. But these are likely to be found elsewhere as exploration of the deep sea progresses. The island slopes and summits of seamounts associated with ocean ridges are important areas for fisheries; evidence suggests that biodiversity, including large pelagic predators, is enhanced around such features.

Polar Deep Sea

Arctic

Arctic deep-sea areas have generally been poorly studied; although several studies over the past two decades have greatly advanced our knowledge of its marine diversity and deep-sea processes. They indicate that the Arctic deep sea is an oligotrophic area, featuring steep gradients in benthic biomass with increasing depth that are primarily driven by food availability.

The Arctic deep basins comprise ~50 per cent of the Arctic Ocean seafloor and differ from those of the North Atlantic, as the Arctic Sea is relatively young in age, semiisolated from the world's oceans, and largely ice-covered. Moreover, the high Arctic experiences more pronounced seasonality in light, and hence in primary production, than lower latitudes.

The history and semi-isolation of the Arctic basin play a major role in its biodiversity patterns. Originally an embayment of the North Pacific, the Arctic deep sea was influenced by Pacific fauna until ~80 million years ago, when the deep-water connection closed. Exchange with the deep Atlantic began ~40 Ma ago, coinciding with a strong cooling period. Although some Arctic shelf and deep-sea fauna were removed by Pleistocene glaciations, other shelf fauna in the Atlantic sector of the Arctic found refuge in the deep sea and are considered the ancestral fauna at least for some of the recent Arctic deep-sea fauna. The bottom of the Arctic basin is filled with water originating from the North Atlantic; the sediments are primarily silt and clay whilst the ridges and plateaus have a higher sand fraction. Exceptions include ice-rafted dropstones, enhancing diversity by providing isolated hard substrata and enhanced habitat heterogeneity for benthic fauna. Considerable inputs of refractory terrestrial organic matter from the large Russian and North American rivers characterize the organic component of sediments along the slopes, and in the basins. The only present-day deep-water connection to the Arctic is via the Fram Strait (~2,500m), providing immigrating species access via the high water flux through this gateway. Submarine ridges within the Arctic form physical barriers, but current evidence suggests that these do not form biogeographic barriers.

Bluhm et al. conservatively estimated the number of benthic invertebrate taxa in the Arctic deep sea to be ~1,125. As in other soft-sediment habitats, foraminiferans and nematodes generally dominate the meiofauna, whereas annelids, crustaceans and bivalves dominate the macrofauna, and echinoderms dominate the megafauna. The degree of endemism at the level of both genera and species is far lower than in the Antarctic, which has a similarly harsh environment. Just over 700 benthic species were catalogued from the central basin a decade ago. The latitudinal species-diversity gradient has been observed in the Arctic Ocean and the peak of the unimodal species-diversity depth gradient occurs at much shallower depths compared to other oceans.

The Arctic is populated by species that have experienced selection pressure for generalism and high vagility, and should have inherent resilience in the face of climate change.

In a warmer future Arctic with less sea ice altered algal abundance and composition will affect zooplankton community structure and subsequently the flux of particulate organic matter to the seafloor, where the changing quantity and quality of this matter will impact benthic communities.

Antarctic

The Southern Ocean comprises three major deep ocean basins, i.e., the Pacific, Indian and Atlantic Basins, separated by submarine ridges and the Scotia Arc island chain. Oceanographically, the Southern Ocean is a major driver of global ocean circulation and plays a vital role in interacting with the deep water circulation in each of the major oceans.

The winter sea-ice formation creates cold, dense, salty water that sinks to the seafloor and forms very dense Antarctic Bottom Water. This in turn pushes the ocean's nutrient-rich, deep water closer to the surface, generating areas of high primary productivity in Antarctic waters, similar to areas of upwelling elsewhere in the world.

The remote Southern Ocean is home to a diverse and rich community of life that thrives in an environment dominated by glaciations and strong currents. However, although relatively little is known about the deep-sea fauna, or about the complex interactions between the highly seasonally variable physical environment and the species that inhabit the Southern Ocean, but our knowledge of Southern Ocean deep-sea fauna and biogeography is increasing rapidly. The range of ecosystems found in each of the marine realms can vary greatly within a small geographic area, or in other cases remain relatively constant across vast areas of the Southern Ocean.

The region also contains many completely un-sampled areas for which nothing is known (e.g., Amundsen Sea, Western Weddell Sea, Eastern Ross Sea). These areas include the majority of the intertidal zone, areas under the floating ice shelves, and the greater benthic part of the deep sea. However, several characteristic features of Southern Ocean ecosystems include circumpolar distributions and eurybathy of many species.

Both pelagic and benthic communities tend to show a high degree of patchiness in both diversity and abundance. The benthic populations show a decrease in biomass with increasing depth, with notable differences in areas of disturbance due to anchor ice and icebergs in the shallows and in highly productive deep fjord ecosystems. Hard and soft sediments from the region are known to be capable of supporting both extremes of diversity and biomass. In some cases, levels of biomass are far higher than those in equivalent habitats in temperate or tropical regions. A major international study led by Brandt revealed comparably high levels of biodiversity (higher than in the Arctic), thereby challenging suggestions that deep-sea diversity is depressed in the Southern Ocean. Understanding of large-scale diversity distributions is improving. For example, depth-diversity gradients of several taxa are known to be unimodal with a shallow peak comparable to those of the Arctic Ocean.

Longline fishing continues in the Southern Ocean, where the Commission for the Conservation of Antarctic Marine Living Resources (CCAMLR) has been implementing conservation measures for toothfish, icefish and krill fisheries, and has closed almost all of the regulatory area to bottom trawling since the 1980s. Climate change, is also a significant potential threat to the Antarctic marine communities, for reasons similar to those presented for the Arctic.

Seamounts

Seamounts are important topographic features of the open ocean. Although they are small in area relative to the vast expanse of the abyssal plains, accounting for < 5 per cent of the seafloor, three important characteristics distinguish them from the surrounding deep-sea habitat. First, they are "islands" of shallow sea floor, and provide a range of depths for different communities. Second, bare rock surfaces can be common, enabling sessile organisms to attach to the rock, in contrast to the majority of the ocean sea floor, which is covered with fine unconsolidated sediments. Third, the physical structure of some seamounts drives the formation of localised hydrographic features and current flows that can keep species and production processes concentrated over the seamount, even increasing the local deep pelagic biomass.

Organic Falls

The decay of large sources of organic matter (e.g., whales, wood, jellyfish) that 'fall' from surface or midwater provide a concentrated source of food on the deep sea floor directly, and indirectly through the decay of the organic matter, can yield hydrogen sulphide and methane. An array of scavenging species (hagfish, amphipods, ophiuroids, and crabs) is adapted to rapidly finding and consuming organic matter on the deep seabed. In addition, lipid-rich whale bones and wood support specialized taxa that have evolutionarily adapted to consume the substrate via symbionts. At least 30 species of polychaetes in the genus Osedax colonize and degrade whale bones, with the aid of heterotrophic symbionts in the group Oceanspirales. Osedax and other taxa colonizing whale falls exhibit biogeographic separation, succession during the life of the whale fall, Adipicola and other deep-sea mussels also harbour chemoautotrophic endosymbionts and colonize sulphide-rich whale remains. Similarly, members of the bivalve genus Xylophaga colonize and consume wood in the deep sea, with symbionts that aid cellulose degradation and nitrogen fixation. The activities of these 'keystone' species, in conjunction with microbial decay, transform the environment and facilitate colonization by a high diversity of other taxa, for example >100 species thus far found only on deep-sea whale falls. Human impacts have likely already affected these organic-fall ecosystems. For example, 20th century whaling drastically reduced the flux of whale carcasses to the deep seafloor.

Numerous areas throughout the world's oceans have experienced large jellyfish population expansions. Although numerous studies have sought to identify the driving forces behind and the impacts of live jellyfish on marine ecosystems, very few have focused on the environmental consequences from the deposition of jellyfish carcasses (from natural die-off events). Recently it has become apparent that jellyfish carcasses have very high sinking speeds. Thus, jellyfish blooms may affect seafloor habitats through the sedimentation of jellyfish carcasses (but also of macrozooplankton,), the smothering of extensive areas of seafloor and reducing oxygen flux into seafloor sediments leading to hypoxic/anoxic conditions. Jelly falls may also be actively consumed by typical deep-sea scavengers, enhancing food-flux into deep-sea food webs. Jellyfish falls have so far been observed in the Atlantic, Indian and Pacific oceans, and are reviewed in Lebrato et al.

Methane Seeps

Continental margins host a vast array of geomorphic environments associated with methane seepage and other types of seeps. Many support assemblages reliant on chemosynthesis fuelled by methane and sulphide oxidation.

Major Ecosystem Services being Affected by the Pressures

Despite its apparent remoteness and in hospitability, the deep ocean and seafloor play a crucial role in human social and economic wellbeing through the ecosystem goods and services they provide. Whilst some services, such as deep-sea fisheries, oil and gas energy resources, potential CO_2 storage, and mineral resources directly benefit humans, other services support the processes that drive deep-sea and global ecosystem functioning. Despite its inaccessibility to most people, the deep sea nonetheless supports important cultural and existence values. The deep sea acts as a sink for anthropogenic CO_2, provides habitat, regenerates nutrients, is a site of primary (including chemosynthetic) and secondary biomass production, as well as providing other biodiversity-related functions and services, including those the deep water and benthic assemblages provide.

Ocean warming and acidification associated with climate change already affect the deep sea, reaching abyssal depths in some areas. Ongoing global climatic changes driven by increasing anthropogenic emissions and subsequent biogeochemical changes portend further impacts for all ocean areas, including the deep-sea and open ocean. Data from preanthropocene times indicates millennial-scale climate variability on deep-sea biodiversity, as well as decadal-centennial climate events.

Some impacts of climate change will be direct. For example, altered distributions and health of open-ocean and deep-sea fisheries are expected to result from warming induced latitudinal or depth shifts; deoxygenation will induce habitat compression; and acidification will stress organismal function and thus organismal distribution. Climate change-related stressors are also likely to act in concert, and effects could be cumulative. Shifts in bottom-up, competitive, or top-down forcing will produce complex and indirect effects on the services described above. Acidification-slowed growth of carbonate skeletons, delayed development under hypoxic conditions, and increased respiratory demands with declining food availability illustrate how climate change could exacerbate anthropogenic impacts and compromise deep-sea ecosystem structure and function and ultimately benefits to human welfare.

The most important ecosystem service of the deep pelagic region is arguably the "biological pump", in which biological processes, such as the daily vertical migration, package and accelerate the transport of carbon compounds, nutrients, and other materials out of surface waters and into the deep sea. However, the microbial diversity and processes of the deep-pelagic realm are not sufficiently known to predict confidently how the biological pump ecosystem service will respond to perturbations.

Deep-sea Exploitation

Deep-sea Fisheries

Deep-sea fishing has a long history, but it did not become an important activity until the mid-twentieth century, when technological advancement allowed the construction of large and powerful vessels and the development of line and trawl gear that could be deployed to continental slope depths. FAO acknowledges that deep-sea fisheries often exploit species which have relatively slower growth rates, reach sexual maturity later and reproduce at lower rates than shelf and coastal species.

Deep-sea fish species were the basis of major commercial fisheries in the 1970s to early 2000s but started to decline as aggregations were fished out, and realisation grew about the low productivity, and hence low yields, of these species and impacts of some of these fisheries on seafloor structure and benthos. Globally the main commercial deep-sea fish species at present number about 20, comprising alfonsino, toothfish, redfish, slickheads, cardinalfish, scabbardfish, armourhead, orange roughy, oreos, roundnose and rough-headed grenadiers, blue ling and moras. The current commercial catch of these main deep-sea species is about 150,000 tons, and has been similar over the last five years, although the proportional species mix has changed.

Deep Gas and Oil Reserves

The oil and gas industry has been active in the open ocean since the 1970s. Over 10,000 hydrocarbon wells have been drilled globally; at least 1,000 are routinely drilled in water depths >200 m, and as deep as 2,896 m in the Gulf of Mexico.

Minerals

Great interest exists in exploiting the deep sea for its various reserves of minerals, which include polymetallic nodules, seafloor massive sulphide (SMS) deposits, mineral-rich sediments and cobalt-rich crusts. Currently no commercial mining projects have started, although several projects are in the exploratory or permitting phase. From those exploratory studies and related research some knowledge of potential ecosystem effects is accumulating.

Experimental studies to assess the potential impact of mining polymetallic nodules in the abyss have indicated that seafloor communities may take many decades before showing signs of recovery from disturbance, and may never recover if they rely directly on the nodules for habitat.

The recovery of communities at active hydrothermal vents where SMS deposits may be exploited may be relatively rapid, because vent sites undergo natural disturbances which have seen some communities appear to recover from catastrophic volcanic activity within a few years. However, the rates of recovery of benthic communities are likely to vary among sites.

Other potential mining activities include exploiting mineral-rich sediments. For example in some deep marine sediments, phosphorite occurs as "nodules" (2 to >150 mm in diameter), in a mud or sand matrix, which can extend beneath the seafloor sediment surface to tens of centimetres depth.

No mining has yet been authorized for such deposits but could result in the removal of large volumes of both the phosphorite nodules and the surrounding soft sediments, together with associated faunal communities and generate large sediment plumes. In addition, cobalt-rich ferromanganese crusts are promising sources of cobalt and rare minerals required to sustain growing human population demands and emerging high and green technologies. Conditions favouring their formation are found in abrupt topography, especially on the flanks and summits of oceanic seamounts and ridges at depths of 800-2500 m, where the most Cobalt-rich deposits are known to concentrate, in habitats dominated by suspension-feeding sessile organisms (mostly cold-water corals and sponges) and comparatively rich biological communities. Interest in cobalt-rich crust resources is growing, although mining for cobalt-rich crusts has not yet started, and technological challenges mean it may develop later than for polymetallic nodule or SMS resources.

Special Conservation/Management Issues and Sustainability for the Future

Special Habitats (VMEs, EBSAs, MPAs) and Conservation Measures

The United Nations General Assembly has adopted a number of resolutions that called for the identification and protection of vulnerable marine ecosystems (VMEs) from significant adverse impacts of bottom fishing (for example 61/105 of 2006), which has facilitated the development of the 2008 International Guidelines for the Management of Deep-Sea Fisheries in the High Seas. Also in the 2000s, in response to the call in the World Summit on Sustainable Development (WSSD) for greater protection of the open ocean, the Conference of Parties to the Convention on Biological Diversity (CBD) developed and adopted criteria for the description of ecologically or biologically significant areas (EBSAs) in open-ocean waters and deep-sea habitats. The application of the EBSA criteria is a scientific and technical exercise, and areas that are described as meeting the criteria may receive protection through a variety of means, according to the choices of States and competent intergovernmental organizations. Expert reviews have concluded that both approaches can be complementary in achieving effective sustainable management in the deep sea.

Protection of the Marine Environment in the Area

With regard to deep-sea mining the International Seabed Authority (ISA), established in 1994, is required to take the necessary measures ensure that the marine environment is protected from harmful effects from activities in the area under its jurisdiction. Such measures may include assessing potential environmental impacts of deep-sea activities (exploration and possible mining) and setting standards for environmental data collection, establishment of environmental baselines, and monitoring programmes.

Deep-ocean Observatories-ocean Networks

Deep-sea observatories are becoming increasingly important in monitoring deep-sea ecosystems and the environmental changes that will affect them. The first long-term and real-time deep-sea observatory was deployed in 1993 at a methane seep site at 1,174 m depth in Sagami Bay, Japan, and is still operating. Several internationally organized projects have been initiated to achieve global integration of deep-sea observatories (e.g., Global Ocean Observing System (GOOS, NSF); FixO3 (Fixed Point Open Ocean Observatories, European Union Framework Programme 7), largely based on existing observing networks (e.g., Porcupine Abyssal Plain in the North Atlantic, (NOC, UK), Hausgarten Site in the transition between the North Atlantic and the Arctic (AWI, Germany), Ocean Network Canada with the Neptune Observatory on Canada's west coast) and aiming at achieving multidisciplinary integration, including physics, climate, biogeochemistry, biodiversity and ecosystems, geophysics with integration across sectors, and economics and sociology. Whilst moving towards a global strategy to obtain maximum efficiency, one of the major goals of deep-sea observatory initiatives is to better understand and predict the effects of climate change on the linked ocean-atmosphere system, and on marine ecosystems, biodiversity and community structure, In terms of biodiversity and ecosystems, several objectives need addressing: exploration and observation; prediction of future biological resources; understanding the functioning of deep-sea ecosystems; and understanding the roles of relationships between ecosystems and the services they provide.

Deep-sea habitats.

Top left: coral garden in the Whittard Canyon, NE Atlantic at approx. 500 metres depth; top right: A sea anemone, Boloceroides daphneae, on cobalt crust covering a seamount off Hawaii, 1000 metres depth; bottom left: An orange roughy (Hoplostethus atlanticus) aggregation at 890 metres depth near the summit of a small seamount (termed "Morgue") off the east coast of New Zealand (image courtesy of Malcolm Clark); bottom right: A reef-like coverage by stony corals (Solenosmilia variabilis) together with prominent orange brisingid seastars on the summit of a small seamount (termed "Ghoul") feature.

Oceanography

Oceanography is also called oceanology or marine science, is the branch of Earth science that studies the ocean. It covers a wide range of topics, including marine organisms and ecosystem dynamics; ocean currents, waves, and geophysical fluid dynamics; plate tectonics and the geology of the sea floor; and fluxes of various chemical substances and physical properties within the ocean and across its boundaries. These diverse topics reflect multiple disciplines that oceanographers blend to further knowledge of the world ocean and understanding of processes within it: biology, chemistry, geology, meteorology, and physics as well as geography.

Traditionally, oceanography has been divided into four separate but related branches: physical oceanography, chemical oceanography, marine geology, and marine ecology. Physical oceanography deals with the properties of seawater (temperature, density, pressure, and so on), its movement (waves, currents, and tides), and the interactions between the ocean waters and the atmosphere. Chemical oceanography has to do with the composition of seawater and the biogeochemical cycles that affect it. Marine geology focuses on the structure, features, and evolution of the ocean basins. Marine ecology, also called biological oceanography, involves the study of the plants and animals of the sea, including life cycles and food production.

Oceanography is the sum of these several branches. Oceanographic research entails the sampling of seawater and marine life for close study, the remote sensing of oceanic processes with aircraft and Earth-orbiting satellites, and the exploration of the seafloor by means of deep-sea drilling and seismic profiling of the terrestrial crust below the ocean bottom. Greater knowledge of the world's oceans enables scientists to more accurately predict, for example, long-term weather and climatic changes and also leads to more efficient exploitation of the Earth's resources. Oceanography also is vital to understanding the effect of pollutants on ocean waters and to the preservation of the quality of the oceans' waters in the face of increasing human demands made on them.

Ocean Exploration

Ocean exploration or undersea exploration is the investigation and description of the ocean waters and the seafloor and of the Earth beneath.

Primary Objectives and Accomplishments

Included in the scope of undersea exploration are the physical and chemical properties of seawater, all manner of life in the sea, and the geological and geophysical features of the Earth's crust. Researchers in the field define and measure such properties; prepare maps in order to identify patterns; and utilize these maps, measurements, and theoretical models to achieve a better grasp of how the Earth works as a whole. This knowledge enables scientists to predict, for example, long-term weather and climatic changes and leads to more efficient exploration and exploitation of the Earth's resources, which in turn result in better management of the environment in general.

The multidisciplinary expedition of the British ship "Challenger" in 1872–76 was the first major undersea survey. Although its main goal was to search for deep-sea life by means of net tows and dredging, the findings of its physical and chemical studies expanded scientific knowledge of temperature and salinity distribution of the open seas. Moreover, depth measurements by wire soundings were carried out all over the globe during the expedition.

Since the time of the "Challenger" voyage, scientists have learned much about the mechanics of the ocean, what it contains, and what lies below its surface. Investigators have produced global maps showing the distribution of surface winds as well as of heat and rainfall, which all work together to drive the ocean in its unceasing motion. They have discovered that storms at the surface can penetrate deep into the ocean and, in fact, cause deep-sea sediments to be rippled and moved. Recent studies also have revealed that storms called eddies occur within the ocean itself and that such a climatic anomaly as El Niño is caused by an interaction of the ocean and the atmosphere.

Other investigations have shown that the ocean absorbs large amounts of carbon dioxide and hence plays a major role in delaying its buildup in the atmosphere. Without the moderating effect of the ocean, the steadily increasing input of carbon dioxide into the atmosphere (due to the extensive burning of coal, oil, and natural gas) would result in the rapid onset of the so-called greenhouse effect—i.e., a warming of the Earth caused by the absorption and reradiating of infrared energy to the terrestrial surface by carbon dioxide and water vapour in the air.

The field of marine biology has benefitted from the development of new sampling methods. Among these, broad ranging acoustical techniques have revealed diverse fish populations and their distribution, while direct, close up observation made possible by deep-sea submersibles has resulted in the discovery of unusual (and unexpected) species and phenomena.

In the area of geology, undersea exploration of the topography of the seafloor and its gravitational and magnetic properties has led to the recognition of global patterns of continental plate motion. These patterns form the basis of the concept of plate tectonics, which synthesized earlier hypotheses of continental drift and seafloor spreading. As noted earlier, this concept not only revolutionized scientific understanding of the Earth's dynamic features (e.g., seismic activity, mountain-building, and volcanism) but also yielded discoveries of economic and political impact. Earth scientists found that the mid-ocean centres of seafloor spreading also are sites of important metal deposits. The hydrothermal circulations associated with these centres produce sizable accumulations of metals important to the world economy, including zinc, copper, lead, silver, and gold. Rich deposits of manganese, cobalt, nickel, and other commercially valuable metals have been found in nodules distributed over the entire ocean floor. The latter discovery proved to be a major factor in the establishment of the Convention of the Law of the Sea, which calls for the sharing of these resources among developed and developing nations alike. Exploitation of these findings awaits only the introduction of commercially viable techniques for deep-sea mining and transportation.

Basic Elements of Undersea Exploration

Platforms

Undersea exploration of any kind must be conducted from platforms, in most cases, ships, buoys, aircraft, or satellites. Typical oceanographic vessels capable of carrying out a full complement of underwater exploratory activities range in size from about 50 to 150 metres. They support scientific crews of 16 to 50 persons and generally permit a full spectrum of interdisciplinary studies. One example of a research vessel of this kind is the "Melville," operated by the Scripps Institution of Oceanography. It has a displacement of 2,075 tons and can carry 25 scientists in addition to 25 crew members. It is powered by a dual cycloidal propulsion system, which provides remarkable manoeuvrability.

The "JOIDES Resolution," operated by Texas A & M University for the Joint Oceanographic Institutions for Deep Earth Sampling, represents a major advance in research vessels. A converted commercial drill ship, it measures 145 metres in length, has a displacement of 18,600 tons, and is equipped with a derrick that extends 62 metres above the waterline. A computer-controlled dynamic positioning system enables the ship to remain over a specific location while drilling in water to depths as great as 8,300 metres. The drilling system of the ship is designed to collect cores from below the ocean floor; it can handle 9,200 metres of drill pipe. The vessel thus can sample most of the ocean floor, including the bottoms of deep ocean basins and trenches. The "JOIDES Resolution" has other notable capabilities. It can operate in waves as high as eight metres, winds up to 23 metres per second, and currents as strong as 1.3 metres per second. It has been outfitted for use in ice so that it can conduct drilling operations in high latitudes. The ship can accommodate 50 scientists as well as the crew and drilling team, and its geophysical laboratories total nearly 930 square metres.

Other specialized vessels include the deep submergence research vehicle known as "Alvin," which can carry a pilot and two scientific observers to a depth of 4,000 metres. The manoeuvrability of the "Alvin" was pivotal to the discoveries of the mineral deposits at the mid-ocean seafloor spreading centres and of previously unknown biological communities living at those sites. Another versatile vessel is the Floating Instrument Platform (FLIP). It is a long narrow platform that is towed in a horizontal position to a research site. Once on location, the ballast tanks are flooded to flip the ship to a vertical position. Only 17 metres of the ship extend above the waterline, with the remaining 92 metres completely submerged. The rise and fall of the waves cause a very small change in the displacement, resulting in a high degree of stability.

New ship designs that promise even greater stability and ease of use include that of the Small Waterplane Area Twin Hull (SWATH) variety. This design type requires the use of twin submerged, streamlined hulls to support a structure that rides above the water surface. The deck shape is entirely unconstrained by the hull shape, as is the case for conventional surface vessels. Ship motion is greatly reduced because of the depth of the submerged hulls. For a given displacement, a SWATH-type vessel can provide twice the amount of deck space that a single-hull ship can, with only 10 percent of the motion of the single-hull design type. In addition, a large centre opening, or well, can be used to display and recover instruments.

Navigation

Exploration of any kind is useful only when the location of the discoveries can be noted precisely. Thus, navigation has always been a key to undersea exploration.

There are various ways by which the position of a vessel at sea can be determined. In cases where external references such as stars or radio and satellite beacons are unavailable or undetectable, inertial navigation, which relies on a stable gyroscope for determining position, is commonly employed. It is far more accurate than the long-used technique of dead reckoning, which is dependent on a knowledge of the ship's original position and the effects of the winds and ocean currents on the vessel.

Another modern position-fixing method is all-weather, long-range radio navigation. It was introduced during World War II as Loran (long-range navigation) A, a system that determines position by measuring the difference in the time of reception of synchronized pulses from widely spaced transmitting stations. The latest version of this system, Loran C, uses low-frequency transmissions and derives its high degree of accuracy from precise time-difference measurements of the pulsed signals and the inherent stability of signal propagation. Users of Loran C are able to identify a position with an accuracy of 0.4 kilometre and a repeatability of 15 metres at a distance of up to about 2,220 kilometres from the reference stations. The Loran C system covers heavily travelled regions in the North Pacific and North Atlantic oceans, parts of the Indian Ocean, and the Mediterranean Sea.

Satellite navigation has proved to be the most accurate method of locating geographical position. A polar-orbiting satellite system called Transit was established in the early 1960s by the United States to provide global coverage for ships at sea. In this system, a vessel pinpoints its position relative to a set of satellites whose orbits are known by measuring the Doppler shift of a received signal—i.e., the change in the frequency of the received signal from that of the transmitted signal. The Transit system suffers from one major drawback. Because of the limited number of system satellites, the frequency with which position determinations can be made each day is relatively low, particularly in the tropics. The system is being improved to provide nearly continuous positioning capability at sea. This expanded version, the Global Positioning System (GPS), is to have 18 satellites, six in each of three orbital planes spaced 120° apart. The GPS is designed to provide fixes anywhere on Earth to an accuracy of 20 metres and a relative accuracy 10 times greater.

Methodology and Instrumentation

Water Sampling for Temperature and Salinity

The temperature, chemical environment, and movement and mixing of seawater are fundamental to understanding the physical, chemical, and biological features of the ocean and the geology of the ocean floor. Traditionally, oceanographers have collected seawater by means of specially adapted water-sampling bottles. The most universal water sampler used today, the Nansen bottle, is a modification of a type developed in the latter part of the 19th century by the Norwegian Arctic explorer and oceanographer Fridtjof Nansen. It is a metal sampler equipped with special closing valves that are actuated when the bottle, attached by one end to a wire that carries it to the desired depth, rotates about that end. A mercury thermometer fastened to the bottle records the temperature at the specified depth. The design of the device is such that, when it is inverted, its mercury column breaks. The amount of mercury remaining in the graduated capillary portion of the thermometer indicates the temperature at the point of inversion. This type of reversing thermometer and the Nansen bottle are extensively used by oceanographers because of their accuracy and dependability in a harsh environment.

The temperature and salinity of the ocean have been mapped with data gathered by many ships over many years. This information is used for tracing heat and water movement and mixing, as well as for making density measurements, which are employed in calculating ocean currents. It was noted as early as the "Challenger" expedition that the salt dissolved in seawater has remarkably constant major constituents. As a consequence, it is possible to map water density patterns within the sea with measurements of only the water temperature and one major property of the sea salt (e.g., the chloride ion content or the electrical conductivity) to arrive at an accurate estimate of the density of a given sample.

Standard laboratory techniques such as titration are routinely used at sea for determining chlorinity. Chlorinity can be briefly defined as the number of grams of chlorine, bromine, and iodine contained in one kilogram of seawater, assuming that the bromine and iodine are replaced by chlorine. Salinity is the total weight of dissolved solids, in grams, found in one kilogram of seawater and may be determined from the concentration of chlorinity because of the constancy of major constituents. In the traditional technique, a solution of silver nitrate of a known strength is added to a sample of seawater to produce the same reaction as with "standard" seawater. The difference in the amounts added gives the degree of chlorinity.

Accurate and continuous measurements of temperature as it changes with depth are required for understanding how the ocean moves and mixes heat. To provide the necessary detail, temperature profilers had to be developed; then, with the introduction of reliable conductivity sensors, salinity profilers were added. An instrument called the bathythermograph (BT), which has been used since the early 1940s to obtain a graphic record of water temperature at various depths, can be lowered from a ship while it is moving at reduced speed. In this instrument a depth element (pressure-operated bellows) drives a slide of smoked glass or metal at right angles to a stylus. Actuated by a thermal element (liquid-filled bourdon tube) that expands and contracts in response to changes in temperature, the stylus scribes a continuous record of temperature and depth.

An expendable bathythermograph (XBT) was developed during the 1970s and has come into increasingly wider use. Unlike the BT, this instrument requires an electrical system aboard the research platform. It detects temperature variations by means of a thermistor (an electrical resistance element made of a semiconductor material) and depends on a known fall rate for depth determination. The sensor unit of the XBT is connected to the research platform by a leak-proof, insulated two-conductor cable. This cable is wound around a pair of large spools in an arrangement resembling that of a fisherman's spinning reel. In operation, the cable is unwound from each of the spools in a direction that is parallel to the axis of the respective spool. As a result, the cable unwinds from both the platform—either a ship or an airplane—and the sensor unit simultaneously but independently. Because of this double-spool arrangement, the sensor unit can free-fall from wherever it hits the sea surface and is completely unaffected by the direction or speed of the craft from which it was deployed. One of the principal reasons why the XBT has proved so useful is that it can provide a record of considerable depth even when it is deployed from a ship moving at full speed.

Until the late 1950s, salinity was universally determined by titration. Since then, shipboard electrical conductivity systems have become widely used. Salinity-Temperature-Depth (STD) and the more recent Conductivity-Temperature-Depth (CTD) systems have greatly improved on-site hydrographic sampling methods. They have enabled oceanographers to learn much about small-scale temperature and salinity distributions.

The most recent version of the CTD systems features rapid-response conductivity and temperature sensors. The conductivity sensor consists of a tiny cell with four platinum electrodes. This type of conductivity cell virtually eliminates errors resulting from the polarization that occurs where the electrodes come in contact with seawater. The temperature sensor combines a tiny thermistor with a platinum-resistance thermometer. Its operations are carried out in such a way as to fully exploit the fast response of the thermistor and the high accuracy of the platinum thermometer. In addition, the system uses a strain gauge as a pressure sensor, the gauge being adjusted to reduce temperature effects to a minimum. This CTD system is extremely reliable. While its temperature precision is greater than 0.001 °C over a range of −3° to +32 °C, its conductivity precision is on the order of one part per million.

Electrical conductivity measurement of seawater salinity has been so effective that it has given rise to a new practical salinity scale, one that is defined on the basis of conductivity ratio. This scale has proved to be a more reliable way of determining density (i.e., the weight of any given volume of seawater at a specified temperature) than the chlorinity scale traditionally used. Such is the case because chlorinity is ion specific while conductivity is sensitive to changes in any ion. Investigators have found that measurements of conductivity ratio make it possible to predict density with a precision almost one order of magnitude greater than was permitted by the chlorinity measurements of the past.

Water Sampling for Chemical Constituents

Nutrient concentration (e.g., phosphate, nitrate, silicate), the pH (acidity), and the proportion of dissolved gases are used by the ocean chemist to determine the age, origin, and movement of water masses and their effect on marine life. Analysis of dissolved gases, for example, is useful in tracing ocean mixing, in studying gas production in the ocean, and in elucidating the natural cycles of atmospheric pollutants. Many such measurements are conducted aboard ship by autoanalyzers, devices that continually monitor a flow of seawater by spectral techniques. Those analyses that cannot be accomplished by an autoanalyzer are carried out with discrete samples in shipboard or shore-based laboratories.

Radioactive chemical tracers are of special interest. Radioisotopes serve as time clocks, thus offering a means of determining the age of water masses, the absolute rates of oceanic mixing, and the generation and destruction of plant tissue. The distribution of these time clocks is controlled by the interaction of physical and biological processes, and so these influences must be disentangled before the clocks can be read. A notable example is the use of carbon-14 (14C). Today, a number of oceanographic laboratories make carbon-14 measurements of oceanic dissolved carbon for the study of mixing and transport processes in the deep ocean. Until recently large samples of water—about 200 litres (one litre = 0.264 gallon)—were required for analysis. New techniques use a linear accelerator (a device that greatly increases the velocity of electrically charged atomic and subatomic particles) as a sophisticated mass spectrometer to directly determine abundancy ratios of carbon-14/carbon-13/carbon-12 atoms. The advantage of the newer methodology is that only very small sample amounts—about 250 millilitres (one millilitre = 0.034 fluid ounce), are required for high accuracy measurements.

Measurements of Ocean Currents

Ocean currents can be measured indirectly through data on density and directly with current meters. In the indirect technique, water density is computed from temperature and salinity observations, and pressure is then calculated from density. The resulting highs and lows of ocean pressure can be used to estimate ocean currents. The indirect technique establishes currents relative to a particular pressure surface; it is best for large-scale, low-frequency currents.

Direct measurement of currents is used to establish absolute currents and to monitor rapidly varying changes. In order to measure currents directly, a current meter must accurately record the speed and direction of flow, and the platform or mooring has to be reliable, readily deployable, and extremely sturdy. Researchers are able to make continuous measurements of currents at levels below the surface layer for periods of more than a year.

A typical system for the direct measure of ocean currents has three principal components: a surface or near-surface float; a line consisting of segments of wire and nylon that holds the current meters; and a release mechanism, signalled acoustically, which will drop an anchor when the system is ready to be brought back. A current meter typically employs a rotor equipped with a small direction vane that moves freely in line with the meter.

One of the most important advances in modern instrument design has been the introduction of low-power, solid-state microelectronics. The accuracy of the Vector Averaging Current Meter (VACM), for example, has been improved appreciably by the use of integrated circuits, as has its data-handling capability. Because of the latter, the VACM can sample the direction and speed of currents roughly eight times during each revolution of the rotor. It then computes the north and east components of speed and stores this data, together with direction and time measurements, on a compact cassette recorder. The VACM is capable of making accurate measurements in wave fields as well as from moorings at the ocean surface because of its direct vector-averaging feature.

Currents also can be measured by drifting floats, either at the surface or at a given depth. Tracking the location of the floats is critical. Surface floats can be followed by satellite, but subsurface drifters must be tracked acoustically. A drifter of this sort acts as an acoustical source and transmits signals that can be followed by a ship with hydrophones suspended into the sea. For such tracking, a low sound frequency is crucial because the higher the frequency of sound, the more rapidly is its energy absorbed by the sea. The longest range floats available during the mid-1980s operated at a frequency as low as about 250 hertz. Long-range floats usually drift along channels known as sound fixing and ranging (SOFAR) channels, which occur in various areas of the ocean where a particular combination of temperature and pressure conditions affect the speed of sound. In a sense, the SOFAR channel acts as a type of acoustic waveguide that focusses sound; as a consequence, several watts of sound can be detected as far away as 2,000 kilometres or so.

Measuring vertical velocity in the ocean posed a major problem for years because of the difficulty of devising a platform that does not move vertically. During the 1960s oceanographers finally came up with a solution, they employed a neutrally buoyant float for measuring vertical velocities. This form of vertical-current meter consists of a cylindrical float on which fins are mounted at an angle. When water moves past the float, it causes the float to turn on its axis. Measurement of the rotation in relation to a compass yields the amount of vertical water movement.

An extension of the neutrally buoyant float is the self-propelled, guided float. One such system, called a Self-Propelled Underwater Research Vehicle (SPURV), manoeuvres below the surface of the sea in response to acoustic signals from the research vessel. It can be used to produce horizontal as well as vertical profiles of various physical properties.

A Doppler-sonar system for measuring upper-ocean current velocity transmits a narrow beam that scatters off drifting plankton and other organisms in the uppermost strata of the ocean. From the Doppler shift of the backscattered sound, the component of water velocity parallel to the beam can be determined to a range of 1,400 metres from the transmitter with a precision of one centimetre per 0.1 second (one centimetre = 0.394 inch).

Integral to a complete picture of the ocean is a profile of velocity. Various methods have been devised for measuring currents as dependent on and varying with depth or horizontal position. Three techniques have been developed to make such measurements: The first involves acoustically tracking a "sinking float" as it descends toward the seafloor. The second technique entails the use of a free-fall device equipped with a current sensor. The third involves a class of current meter specially designed to move up and down a fixed line attached to a vessel, mooring, or drifting buoy. One such instrument has a roller block that couples the front of the instrument to a wire from the vessel. In this way, the motion of the vessel is decoupled from that of the instrument. Another important component of this instrument is its hull, a structure that not only furnishes buoyancy but also serves as a direction vane. In the bottom of the hull is a device that records velocity, temperature, and depth. The entire system descends at a rate of approximately 10 centimetres per second, resulting in a vertical resolution of several metres for the velocity profile produced.

Acoustic and Satellite Sensing

Remote sensing of the ocean can be done by aircraft and Earth-orbiting satellites or by sending acoustic signals through it. These techniques all offer a more sweeping view of the ocean than can be provided by slow-moving ships and hence have become increasingly important in oceanographic research.

Satellite-borne radar altimeters have proved to be especially useful. A radar system of this type can determine the distance between the satellite and the sea surface to an accuracy of better than 10 centimetres by measuring the time it takes for a transmitted pulse of radio energy to travel to the surface and return. By combining such a precise distance measurement with information about the satellite's orbit, oceanographers are able to produce maps of sea-surface topography. Moreover, they can deduce the pressure field of the sea surface by combining the distance measurement with knowledge about the geoid. They can in turn extrapolate information about the general circulation of the upper stratum of the ocean from a synoptic view of the surface pressure field.

Another remote-sensing technique involves the use of satellite-borne infrared and microwave radiometers to measure radiant energy released from the surface of the ocean. Such measurements are used to determine sea-surface temperature. High-resolution, infrared images transmitted by polar-orbiting satellites have provided researchers with an effective means of monitoring wave features in ocean currents over a wide area, as, for example, long equatorial waves in the Pacific Ocean and time variations in the flow of the Gulf Stream between Florida and Cape Hatteras, North Carolina.

Acoustic techniques also have many applications in the study of the ocean, particularly of those subsurface processes and physical properties inaccessible to satellite observation. In one such technique, the temperature structure of a water column from a given point on the seafloor to the surface is studied using an inverted echo sounder. This instrument, which features both an acoustic transmitter and a receiver, measures the time taken by a pulse of sound to travel from the sea bottom to the surface and back again. In most cases, a change in the average temperature of the water column above the instrument causes a fluctuation in the time interval between the transmission and the reception of the acoustic signal.

Other acoustic techniques can be utilized to study ocean variables on a large scale. A method known as ocean acoustic tomography, for example, monitors the travel time of sound pulses with an array of echo-sounding systems. In general, the amount of data collected is directly proportional to the product of the number of transmitters and receivers, so that much information on averaged oceanic properties can be gathered within a short period of time at relatively low cost.

Collection of Biological Samples

Life at the bottom, benthos, is affected by the water column and by the sediment–water interface; the swimmers, or nekton, are influenced by the water that they come in contact with; and the floaters, or plankton—phytoplankton (plant forms) and zooplankton (animal forms)—are influenced by the water and the transfers that occur at the surface of the sea. Thus, in most cases, measurements and sampling of marine life is best done in concert with measurements of the physical and chemical properties of the ocean and the surface effects of the atmosphere.

As a consequence of the close interaction of sea life and its environment, marine biologists and biological oceanographers use most of the techniques mentioned above as well as some specialized techniques for biological sampling. Investigative techniques include the use of sampling devices, remote sensing of surface life-forms by satellite and aircraft, and in situ observation of plants and animals in direct interaction with their environment. The latter is becoming increasingly important as biologists recognize the fragility of organisms and the difficulty of obtaining representative samples. The absence of good sampling techniques means that even today little is known about the distribution, number, and life cycles of many of the important species of marine life.

Some of the most commonly used samplers are plankton nets and midwater trawls. Nets have a mesh size smaller than the plankton under investigation; trawls filter out only the larger forms. The smaller net sizes can be used only when the ship is either stopped or moving ahead slowly; the larger can be used while the ship is travelling at normal speeds. Plankton nets can be used to sample at one or more depths. Qualitative samplers sieve organisms from the water without measuring the volume of water passing through, whereas quantitative samplers measure the volume and hence the concentration of organisms in a unit volume of seawater.

The Clark-Bumpus sampler is a quantitative type designed to take an uncontaminated sample from any desired depth while simultaneously estimating the filtered volume of seawater. It is equipped with a flow meter that monitors the volume of seawater that passes through the net. A shutter opens and closes on demand from the surface, admitting water and spinning the impeller of the meter while catching the plankton. When the impeller is stopped by closing the shutter, the sampler can be raised without contamination from plankton in the waters above.

The midwater trawl is specially designed for rapid collection at depths well below the surface and at such a speed that active, fast-swimming fish are unable to escape from the net once caught. Trawls can be towed at speeds up to nine kilometres per hour. To counteract the tendency of an ordinary net to surface behind the towing vessel, a midwater trawl of the Isaacs-Kidd variety uses an inclined-plane surface rigged in front of the net entrance to act as a depressor. The trawl is shaped like an asymmetrical cone with a pentagonal mouth opening and a round closed end. Within the net, additional netting is attached as lining. A steel ring is fastened at the end of the net to maintain shape. A large perforated can is fastened by drawstrings on the end of the net to retain the sample in relatively undamaged condition.

The use of acoustics to record and measure the distribution of biological organisms is becoming a widely adopted practice. Some organisms can be tracked directly by their distinctive sounds. By recording and analyzing these sounds, biologists are able to chart the behaviour and distribution of such life-forms.

Organisms that passively affect various electronic systems are large mammals, schools of fish, and plankton that either scatter sound or so appear as false targets or background reverberation, or that attenuate the acoustic signal. Some fishes and invertebrates make up layers of acoustic-scattering material, which may exhibit daily vertical movement related to daily changes in light.

Light in the upper layers of the ocean is crucial to maintaining marine life. The penetration and absorption of light and the colour and transparency of the ocean water are indicative of biological activity and of suspended material. In situ measurements of water transparency and absorption include the submarine photometer, the hydrophotometer, and the Secchi disk. The submarine photometer records directly to depths of about 150 metres the infrared, visible, and ultraviolet portions of the spectrum. The hydrophotometer has a self-contained light source that allows greater latitude in observation because it can be used at any time of night or day and measures finer gradations of transparency. The Secchi disk, designed to measure water transparency, is a circular white disk that is lowered on a cable into the sea. In practice, the depth at which it is barely visible is noted. The greater the depth reading, the more transparent is the water.

The primary productivity of the ocean, which occurs in the upper layers, can be monitored by continuous measurement of absorption by chlorophyll molecules. This occurs in the red and blue portions of the spectrum, leaving the green to represent the characteristic colour of biological activity. Satellite measurements of ocean colour that span a number of wavelengths in the visible and infrared portions of the spectrum are used to give a large-scale view of the biological activity and suspended material in the ocean.

Exploration of the Seafloor and the Earth's Crust

The ocean floor has the same general character as the land areas of the world: mountains, plains, channels, canyons, exposed rocks, and sediment-covered areas. The lack of weathering and erosion in most areas, however, allows geological processes to be seen more clearly on the seafloor than on land. Undisturbed sediments, for example, contain a historical record of past climates and the state of the ocean, which has enabled geologists to find a close relation between past climates and the variation of the distance of the Earth from the Sun (the Milankovich effect).

Because electromagnetic radiation cannot penetrate any significant distance into the sea, the oceanographer uses acoustic signals, explosives, and earthquakes, as well as gravity and magnetic fields, to probe the seafloor and the structure beneath. Such techniques—which now include the capability to produce a swath, or two-dimensional, description of the seafloor beneath a ship—are providing increasingly accurate data on the shape of the ocean, its roughness, and the structure beneath. Satellite techniques are a more recent development. Because the shape of the sea surface is closely related to that of the seafloor due to gravity, satellite measurements of surface topography have been used to provide a global view of the ocean bottom. They also have provided data for an accurate mapping of such features as seamounts.

Research on marine sedimentation involves the study of deposition, composition, and classification of organic and inorganic materials found on the seafloor. Samples of such materials are thoroughly examined aboard research vessels or in shore-based laboratories, where investigators analyze the size and shape of constituent particles, determine chemical properties such as pH, and identify and categorize the minerals and organisms present. From thousands of reported classifications and collected samples, bottom-sediment charts are prepared.

Various kinds of equipment are used to obtain samples from the seafloor: These include grabbing devices, dredges, and coring devices. Grabbing devices, commonly known as snappers, vary widely in size and design. One general class of such devices is the clamshell snapper, which is used to obtain small samples of the superficial layers of bottom sediments. Clamshell snappers come in two basic varieties. One measures 76 centimetres in length, weighs roughly 27 kilograms (one kilogram = 2.2 pounds), and is constructed of stainless steel. The jaws of this device are closed by heavy arms, which are actuated by a strong spring and lead weight. It is capable of trapping about a pint of bottom material. The second type of clamshell snapper is appreciably smaller. Commonly called the mud snapper, this device is approximately 28 centimetres long and weighs 1.4 kilograms. Other grabbing devices include the orange peel bucket sampler, which is used for collecting bottom materials in shallow waters. A small hook attached to the end of the lowering wire supports the sampler as it is lowered and also holds the jaws open. When contact is made with the bottom, the sampler jaws sink into the sediment and the wire tension is released, allowing the hook to swing free of the sampler. Upon hoisting, the wire takes a strain on the closing line, which closes the jaws and traps a sample. The underway bottom sampler, or scoopfish, is designed to sample rapidly without stopping the ship. It is lowered to depths less than 200 metres from a ship moving at speeds no more than 28 kilometres per hour. The sampler weighs five kilograms and can capture samples ranging from mud to coral.

The second major category of bottom sampler is the dredge, which is dragged along the seafloor to collect materials. Bottom-dredging operations require very sturdy gear, particularly when dredging for rock samples. A typical dredge is constructed of steel plate and is 30 centimetres deep, 60 centimetres wide and 90 centimetres long. The forward end is open, but the aft end has a heavy grill of round steel bars that is designed to retain large rock samples. When finer sized material is sought, a screen of heavy hardware cloth is placed over the grill.

Coring devices typically have three principal components: interchangeable core tubes, a main body of streamlined lead weights, and a tailfin assembly that directs the corer in a vertical line to the ocean bottom. The amount of sediment collected depends on the length of the corer, the size of the main weight, and the penetrability of the bottom. One type of coring device, the lightweight Phleger corer, takes samples only of the upper layer of the ocean bottom to a depth of about one metre. Deeper cores are taken by the piston corer. In this device, a closely fitted piston attached to the end of the lowering cable is installed inside the coring tube. When the coring tube is driven into the ocean floor, friction exerts a downward pull on the core sample. The hydrostatic pressure on the ocean bottom, however, exerts an upward pressure on the core that will work against a vacuum being created between the piston and the top of the core. The piston, in effect, provides a suction that overcomes the frictional forces acting between the sediment sample and the inside of the coring tube. The complete assembly of a typical piston core weighs about 180 kilograms and can be used to obtain samples as long as 20 metres. An improved version of this device, the hydraulic piston corer, is used by deep-sea drilling ships such as the "JOIDES Resolution." Essentially undisturbed cores of lengths up to 200 metres have been obtained with this type of corer.

Investigators ay also make use of wire-line logging tools that are capable of measuring electrical resistance, acoustic properties, and magnetic and gravitational effects in the holes drilled. The "JOIDES Resolution" is equipped with tools of this sort, including a remote television camera, which are lowered into a drill hole after the core has been removed. Such wire-line logging apparatus make data immediately available for scientific analysis and decision making.

Acoustic techniques have reached a high level of sophistication for geological and geophysical studies. Such multifrequency techniques as those that employ Seabeam and Gloria (Geological Long-Range Inclined Asdic) permit mapping two-dimensional swaths with great accuracy from a single ship. These methods are widely used to ascertain the major features of the seafloor. The Gloria system, for example, can produce a picture of the morphology of a region at a rate of up to 1,000 square kilometres per hour. Techniques of this kind are employed in conjunction with seismic reflection techniques, which involve the use of multichannel receiving arrays to detect sound waves triggered by explosive shots (e.g., dynamite blasts) that are reflected off of interfaces separating rocks of different physical properties. Such techniques make it possible to measure the structure of the Earth's crust deep below the seafloor.

References

- Ocean, entry: newworldencyclopedia.org, Retrieved 31 March, 2019

- Open-ocean-deep-sea: researchgate.net, Retrieved 14 July, 2019

- Oceanography, definition: definitions.net, Retrieved 17 May, 2019

- Oceanography, science: britannica.com, Retrieved 19 April, 2019

- Undersea-exploration, technology: britannica.com, Retrieved 5 February, 2019

Chapter 2

Branches of Oceanography

Oceanography can be further divided into several branches. These include physical oceanography, biological oceanography, chemical oceanography, paleoceanography and geologic oceanography. All these branches have been explored in detail in this chapter.

Biological Oceanography

Biological oceanography is the study of life in the oceans—the distribution, abundance, and production of marine species along with the processes that govern species' spread and development.

It involves:

- Chemical and physical factors influencing distribution patterns.

- Physiological, behavioral, and biochemical adaptations to environmental variables, including natural variations in food, temperature, pressure, light, and the chemical environment.

- Food chain dynamics.

- Nutrient cycling and initial steps of chemical energy fixation.

- Responses to the results of man's activities in the oceans.

Biological oceanography is also a study in extremes:

- In size, from the tiny microbes in the water column to the 30-meter blue whale.

- In depth, from blooms of cyanobacteria covering thousands of square kilometers of the ocean's surface, to hydrothermal vent colonies emerging eerily from the mile-deep dark.

- In locale, from the lab next door to the deck of a research vessel bucking ice flows in the Arctic.

Extreme Habitats

Studying organisms that occupy extreme habitats, such as the deep sea or Polar Regions, presents major challenges for accessibility and sampling. Meeting these challenges has had exciting results. The discovery of biological communities near hydrothermal vents, for example, led to new perceptions about the evolution of life, the discovery of unusual symbioses, and new insights on adaptation in the deep sea.

Tools and Technology

Biological oceanographers rely on a variety of tools and use a variety of approaches to aid them in their study of life in the sea. Some studies involve laboratory experiments with individual organisms. In other cases, the oceanographer must go into the water to directly sample and observe certain types of organisms such as zooplankton.

Other approaches involve underwater submersible vehicles to gain access to biological communities deep in the ocean, such as those associated with deep-sea hydrothermal vents. Many oceanographers use research vessels from which they lower instruments and specialized water sampling gear into the water. Biological oceanographers employ methods derived from various fields, including molecular biology, immunology, physiology, biochemistry, ecology, and many others.

In addition to making scientific observations, the biological oceanographer uses a variety of models to study the biology of the oceans. Theoretical models are used to examine problems in biological oceanography that cannot be answered through direct observation and measurement.

This oceanography graduate student repairs a deep-water instrument in
the biological research lab aboard a research vessel. Deployment of such
devices allows biological oceanographers to conduct measurements at depth.

Heuristic Models are used to help to understand and explain an existing set of observations. Finally, some models are used to predict changes in biological processes that may occur because of natural and human-induced changes to the ocean environment.

Advances in technology have given biological oceanographers new insights about the living oceans. Lasers, fiber optics, high-speed digital video imaging and DNA microarrays are some of the high-tech "gadgets" that are used to study biological processes in the oceans. Robotic underwater vehicles reduce the risk and expense of manned submersibles while providing spectacular views of undersea communities. Other types of instruments are allowed to drift freely with ocean currents, towed behind a ship, or anchored at specific locations to provide detailed information over time and space. Among the most powerful tools available to biological oceanographers are satellite and airborne sensors, which provide large-scale views of the ocean and have greatly enriched the scientific understanding of biological processes and their relationship to physical phenomena.

Major research programs in biological oceanography examine cycles of carbon and other biologically critical elements, such as nitrogen, phosphorus, silicon and iron. These biogeochemical cycles are key in understanding large-scale phenomena such as global warming. Living organisms, particularly phytoplankton (single-celled microscopic plants that utilize photosynthesis), bacterio plankton (marine bacteria), and small animals (zooplankton), play a critical role in biogeochemical cycles.

Other important areas of study include understanding linkages between different levels of the marine food web, from phytoplankton all the way up to fish and marine mammals. Biological oceanographers also study factors that influence biological diversity within the oceans, and the importance of diversity in maintaining biological function. Understanding and mitigating the decline in biodiversity such as has occurred with losses of highly diverse coral reef communities, is a primary concern of biological oceanographers. Researchers also may deal with issues that affect society such as water pollution, overexploitation of fisheries, and harmful algal blooms.

Funding Sources

Biological oceanographers compete for a limited pool of funds to do their research by submitting proposals or bidding on contracts to various scientific agencies. Research is supported by federal agencies such as the National Science Foundation, Department of Commerce, Department of Defense, National Aeronautics and Space Administration, Environmental Protection Agency, Department of Energy, Minerals Management Service, and National Research Council, as well as many other government and private agencies. Funded programs strive to advance a basic understanding of the oceans and life within, provide strategic information required for national defense, and preserve and protect the valuable resources of the oceans.

Chemical Oceanography

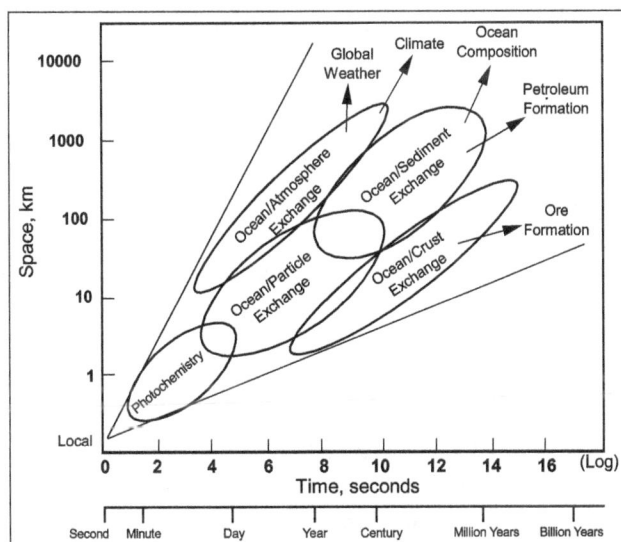

Chemical oceanography is the study of everything about the chemistry of the ocean based on the distribution and dynamics of elements, isotopes, atoms and molecules. This ranges from fundamental physical, thermodynamic and kinetic chemistry to two-way interactions of ocean chemistry with biological, geological and physical processes. It encompasses both inorganic and organic chemistry, and includes studies of atmospheric and terrestrial processes as well. Chemical oceanography includes processes that occur on a wide range of spatial scales; from global to regional to local to microscopic dimensions, and temporal scales; from geological epochs to glacial-interglacial to millennial, decadal, interannual, seasonal, diurnal and all the way to microsecond time scales. The field by its own nature is very much an interdisciplinary field.

The advantages of the chemical perspective include:

1. Huge information potential due to large number of elements (93), isotopes (260), naturally occurring radioisotopes (78) and compounds (innumerable) present in the ocean.

2. Chemical measurements in the ocean are highly representative, reproducible and predictable (statistically meaningful). One drop of water is about $1/20^{th}$ of a milliliter or 0.05 g, this is 2.8×10^{-3} moles or 1.7×10^{21} molecules.

3. Quantitative treatments are possible (stoichiometry's, balances, predictions of reaction rates and extents).

4. We can learn about processes from chemical changes. Seawater composition integrates multiple previous events, this is important because most of the ocean is inaccessible to direct observation.

Ocean Biogeochemistry

Biogeochemistry is the study of the interactions of the biology, chemistry, and geology of the Earth. In the case of a large body of water such as the ocean, biogeochemistry can be thought of as a huge experiment or set of reactions. Instead of happening in a clean glass beaker, the reactions have the ocean floor as the container.

The surface of the water is open to the air, and every day more dust and dirt from land blows over the ocean and falls in. Moreover, the surface of the water contains many small plant forms that are continually growing and being consumed by animals that are themselves consumed by other animals.

As this life and death drama continues, the scraps and leftovers drift downward towards the ocean floor like a snowfall; hence the name "marine snow." Around the edges of the ocean, rivers empty water and sediment. Deep in the ocean, mud-dwelling creatures await the arrival of their next meal from the falling biological debris (marine snow). These events are linked to each other, to the history of life on Earth, and to variations in Earth's climate.

Cycles

Scientists who study biogeochemistry usually consider the cycling of materials through the different parts of the system. To do this, they deal with reservoirs of materials and the fluxes of a substance from one reservoir to another. For example, they examine reservoirs such as the surface ocean water versus the deep ocean water, or the transfer of masses of materials per unit time (fluxes).

Most scientific study has focussed on the carbon cycle. Carbon, after all, is the basis of life on Earth, and its gaseous form, carbon dioxide, is linked to the greenhouse effect and changes in Earth's climate over time. For these reasons, understanding the carbon cycle has been the focus of several large research programs supported by the U.S. government. Three examples include:

- The U.S. Global Change Research Program (USGCRP): A joint project to design a carbon cycle research program; funded by the Department of Energy; the National Aeronautic and Space Administration, the National Oceanic and Atmospheric Administration, National Science Foundation, and U.S. Geological Survey;

- Global Ocean Ecosystems Dynamics (GLOBEC): A major research program funded by the National Science Foundation to determine how global change affects the marine ecosystem and what the feedbacks to the physical climate system will be; and

- The Global Carbon Program (GCP): A study funded by the National Oceanic and Atmospheric Administration to improve scientists' ability to predict the fate of human-derived carbon dioxide and future concentrations of atmospheric carbon dioxide.

Other substances also have well-studied cycles. Water, of course, is constantly moving into, though, and out of the ocean. Some of the atmospheric gases such as oxygen and carbon dioxide are vitally important to life. Nutrient elements such as nitrogen, phosphorus, and silicon are necessary to the phytoplankton, and form the basis for the oceanic food web.

A Cycling Example

The presence of life forms on Earth is tremendously important in the cycling of elements through the major reservoirs. Consider the ocean as an example: If one focuses on the impact of a single diatom on the ocean, the following story emerges.

Diatoms are a group of algae living by the millions in each cubic centimeter of surface ocean water. There each alga has access to the sunlight needed for photosynthesis; the CO_2 (carbon dioxide), N (nitrogen), and P (phosphorus) needed to make its soft tissue; the Si (silicon) needed for its shell-like covering; and a number of rare or trace substances in sea water, including Cu (copper) and Fe (iron). To reproduce, it undergoes cell division. Its life processes produce O_2 (oxygen) that can be used by other organisms; organic tissue that becomes food for the next higher creatures in the food web; and often an exudate or slime.

Once the diatom has been consumed by an animal (a copepod, for example), its life is over, but its effect on the ocean is not. The copepod digests and derives energy from the diatom's soft tissue, then packages the remains into a fecal pellet that is discharged as waste to become part of the falling debris (marine snow) headed for the ocean floor.

The pellet lands on the ocean floor, forming a site for bacteria to live as well as food for them to consume. The inorganic part of the diatom that remains (the silica shell) will begin to dissolve on the way to the ocean bottom, and Si taken out of the surface water is returned to deeper water as the shell dissolves. Decomposition of sinking organic matter by bacteria returns N, C, and P to the water and removes dissolved O_2.

Carbon

Ocean water itself is changed by life processes. During the growth of diatoms and the consumption of diatoms by zooplankton, carbon is removed from ocean water and in turn from the atmosphere as the diatoms use it to grow. The transfer of this carbon toward the ocean floor and its partial burial in the sediments is often referred to as the carbon pump; it is one of the processes that slow the accumulation of CO_2 in the atmosphere.

Silicon

The silicon (Si) used in the diatom shell enters the ocean from rivers, from the hot springs along mid-ocean ridges and by diffusion from deep-sea sediments. Diatoms remove Si so efficiently from the ocean surface water that it is a very scarce element there, and mixing and upwelling processes are necessary to redistribute enough Si back to the surface to provide for diatom growth. For that reason, Si as well as N, P, and other biologically important elements are in low concentration in surface water of the ocean, and increase with depth.

Oxygen

Another consequence of ocean biogeochemistry can be seen in the distribution of O_2 (oxygen) with depth. The oxygen content at the surface is relatively high (about 6 milliliters per liter) and is replenished from the air. Deeper in the water, the O_2 content begins to decrease with depth, until at about 1,000 meters (3,082 feet), the value reaches a minimum. The reason for the decrease is the consumption by bacteria of the rain of organic debris (marine snow) falling through the water. The process requires O_2, and below the surface there is no immediate source to return the O_2 being used up.

The exact amount of O_2 at the O_2 minimum varies with location in the ocean; below the minimum, O_2 content begins to increase again with depth. The increase is related to water circulation in the ocean. The deep water in the ocean starts out at the surface in polar regions, where it becomes very dense because of the extreme cold, and sinks to great depths in the ocean, carrying with it dissolved oxygen from the surface waters. This cold, dense, deep water flows along the ocean floor close to the bottom, well beneath the depths of the O_2 minimum. These factors combine to give the observed shapes of O_2 profiles in the ocean.

Hydrothermal Processes

There are other processes that play a role in determining the nature of the ocean. For example, hydrothermal activity at mid-ocean ridges results in significant changes in the chemistry of ocean water. The water that comes out of these hot springs comes from normal deep-ocean water that runs down into deep cracks on the ocean floor alongside the ridges. As the water penetrates into the oceanic crust, it becomes heated to very high temperatures, and reacts with the rocks. The water that comes out of the vents is very hot; contains sulfide (S^-) instead of sulfate (SO_4^{2-}); contains no Mg (magnesium) or O_2 (oxygen); and contains large amounts of Si (silicon). Because the entire volume of the ocean circulates through the mid-ocean ridge system every 10 million years, these changes are of great significance to the oceans and the organisms that live in them.

Ocean Acidification

Ocean acidification is the worldwide reduction in the pH of seawater as a consequence of the absorption of large amounts of carbon dioxide (CO_2) by the oceans. Ocean acidification is largely the result of loading Earth's atmosphere with large quantities of CO_2, produced by vehicles and industrial and agricultural processes. Since the beginning of the Industrial Revolution about 1750, roughly one-third to one-half of the CO_2 released into Earth's atmosphere by human activities has been absorbed by the oceans. During that time period, scientists have estimated, the average pH of seawater declined from 8.19 to 8.05, which corresponds to a 30 percent increase in acidity.

Some scientists estimate that the pace of ocean acidification since the beginning of the Industrial Revolution has been approximately 100 times more rapid than at any other time during the most recent 650,000 years. They note that concentrations of atmospheric CO_2 between 1000 and 1900 CE ranged between 275 and 290 parts per million by volume (ppmv). In 2010 the average concentration was 390 ppmv, and climatologists expect the concentration to rise to between 413 and 750 ppmv by 2100, depending on the level of greenhouse gas emissions. With additional CO_2 transferred to the oceans, pH would decline further; under worst-case scenarios, seawater pH would drop to between 7.8 and 7.9 by 2100.

Marine scientists are concerned that the process of ocean acidification constitutes a threat to sea life and to the cultures that depend on the ocean for their food and livelihood. Increases in ocean acidity reduce the concentration of carbonate ions and the availability of aragonite (a significant source of calcium carbonate) in seawater. Marine scientists expect that coral, shellfish, and other marine calcifiers (that is, organisms that use carbonates) will be less able to obtain the raw materials that they use to build and maintain their skeletons and shells. These scientists also note that rising ocean acidity presents a number of other physiological problems to different groups of marine organisms and those problems could threaten the stability of marine food chains.

Changes in Seawater Chemistry

The acidity of any solution is determined by the relative concentration of hydrogen ions (H^+). A larger concentration of H^+ ions in a solution corresponds to higher acidity, which is measured as a lower pH. When CO_2 dissolves in seawater, it creates carbonic acid (H_2CO_3) and liberates H^+, which subsequently reacts with carbonate ions (CO_3^{2-}) and aragonite (the stable form of calcium carbonate) to form bicarbonate (HCO_3^-). At present seawater is extremely rich in dissolved carbonate minerals. However, as ocean acidity increases, carbonate ion concentrations fall.

The absorption of CO_2 largely results from the dissolution of the gas into the upper layers of the ocean, but CO_2 is also brought into the oceans through photosynthesis and respiration. Algae and other marine photosynthesizers take in CO_2 and store it in their tissues as carbon. Carbon is then passed to zooplankton and other organisms through the food chain, and these organisms can release CO_2 to the oceans through respiration. In addition, when marine organisms die and fall to the ocean floor, CO_2 is released through the process of decomposition.

Physiological and Ecological Effects

Under the worst-case scenarios outlined above, with seawater pH dropping to between 7.8 and 7.9, carbonate ion concentrations would decrease by at least 50 percent as acids in the seawater reacted with them. Under such conditions, marine calcifiers would have substantially less material to maintain their shells and skeletons. Laboratory experiments in which the pH of seawater has been lowered to approximately 7.8 (to simulate one projected oceanic pH for the year 2100) have shown that such organisms placed in these environments do not grow as well as those placed in environments characterized by early 21st-century levels of seawater acidity (pH = 8.05). As a result, their small size places them at higher risk of being eaten by predators. Furthermore, the shells of some organisms—for instance, pteropods, which serve as food for krill and whales—dissolve substantially after only six weeks in such high-acid environments.

Sea butterfly: The sea butterfly (Limacina helicina),
a pteropod mollusk, displaying a thin outer shell made
transparent by increased acidity in Earth's oceans.

Larger animals such as squid and fishes may also feel the effects of increasing acidity as carbonic acid concentrations rise in their body fluids. This condition, called acidosis, may cause problems with the animal's respiration as well as with growth and reproduction.

In addition, many marine scientists suspect the substantial decline in oyster beds along the West Coast of the United States since 2005 to be caused by the increased stress ocean acidification places on oyster larvae. (It may make them more vulnerable to disease).

Physiological changes brought on by increasing acidity have the potential to alter predator-prey relationships. Some experiments have shown that the carbonate skeletons of sea urchin larvae are smaller under conditions of increased acidity; such a decline in overall size could make them more palatable to predators who would avoid them under normal conditions. In turn, decreases in the abundance of pteropods, foraminiferans, and coccoliths would force those animals that consume them to switch to other prey. The process of switching to new food sources would cause several predator populations to decline while also placing predation pressure on organisms unaccustomed to such attention.

Many scientists worry that many marine species, some critical to the proper functioning of marine food chains, will become extinct if the pace of ocean acidification continues, because they will not have sufficient time to adapt to the changes in seawater chemistry. The world's coral reefs, which provide habitat to many species and are often regarded by ecologists as centres of biodiversity in the oceans, could decline and even disappear if ocean acidification intensifies and carbonate ion concentrations continue to fall.

The deeper waters of the ocean are naturally more acidic than the upper layers, since CO_2 that dissolves at the surface descends with dense, cold water as part of the thermohaline circulation.

The acidic lower layers of the ocean are separated from the upper layers by a boundary called the "saturation horizon." Above this boundary there are enough carbonates present in the water to support coral communities. In midlatitude waters and in waters closer to the poles, many so-called cold-water coral communities are found at depths that range from 40 to 1,000 metres (about 130 to 3,300 feet)—as opposed to their warm-water counterparts, the tropical coral reefs, which are rarely found below 100 metres (330 feet). Since about the year 1800, studies have shown, increased acidity has raised the saturation horizon about 50 to 200 metres (about 160 to 660 feet) in midlatitude and polar waters. This change is enough to threaten cold-water coral communities, and some scientists fear that additional communities will be placed at risk if the boundary approaches the surface of the ocean. A decline in cold-water marine calcifiers would result in a decline in reef building, and other marine organisms that depend on corals for their habitat and food would decline as well. Scientists also predict that, if ocean acidification were to increase worldwide, warm-water coral communities, which often supply food and tourism revenue to people who live near them, would suffer similar fates.

A diver exploring a coral reef in the Maldives.

In addition, scientists predict that the reduction of marine phytoplankton populations due to rising pH levels in the oceans will produce a positive feedback that intensifies global warming. Marine phytoplankton produce dimethyl sulfide (DMS), a gas that serves as the most significant source of sulfur in Earth's atmosphere. Sulfur in Earth's upper atmosphere reflects some of the incoming solar radiation back into space and thus keeps it from warming Earth's surface. Models predict that DMS production will decrease by about 18 percent by 2100 from preindustrial levels, which will result in additional radiative forcing corresponding to an atmospheric temperature increase of 0.25 °C (0.45 °F).

Impacts on Ocean Life

The pH of the ocean fluctuates within limits as a result of natural processes, and ocean organisms are well-adapted to survive the changes that they normally experience. Some marine species may be able to adapt to more extreme changes—but many will suffer, and there will likely be extinctions. We can't know this for sure, but during the last great acidification event 55 million years ago, there were mass extinctions in some species including deep sea invertebrates. A more acidic ocean won't destroy all marine life in the sea, but the rise in seawater acidity of 30 percent that we have already seen is already affecting some ocean organisms.

Coral Reefs

Branching corals, because of their more fragile structure,
struggle to live in acidified waters around natural carbon
dioxide seeps, a model for a more acidic future ocean.

Reef-building corals craft their own homes from calcium carbonate, forming complex reefs that house the coral animals themselves and provide habitat for many other organisms. Acidification may limit coral growth by corroding pre-existing coral skeletons while simultaneously slowing the growth of new ones, and the weaker reefs that result will be more vulnerable to erosion. This erosion will come not only from storm waves, but also from animals that drill into or eat coral. A recent study predicts that by roughly 2080 ocean conditions will be so acidic that even otherwise healthy coral reefs will be eroding more quickly than they can rebuild.

Acidification may also impact corals before they even begin constructing their homes. The eggs and larvae of only a few coral species have been studied, and more acidic water didn't hurt their development while they were still in the plankton. However, larvae in acidic water had more trouble finding a good place to settle, preventing them from reaching adulthood.

Some types of coral can use bicarbonate instead of carbonate ions to build their skeletons, which gives them more options in an acidifying ocean. Some can survive without a skeleton and return to normal skeleton-building activities once the water returns to a more comfortable pH. Others can handle a wider pH range.

Nonetheless, in the next century we will see the common types of coral found in reefs shifting—though we can't be entirely certain what that change will look like. On reefs in Papua New Guinea that are affected by natural carbon dioxide seeps, big boulder colonies have taken over and the delicately branching forms have disappeared, probably because their thin branches are more susceptible to dissolving. This change is also likely to affect the many thousands of organisms that live among the coral, including those that people fish and eat, in unpredictable ways. In addition, acidification gets piled on top of all the other stresses that reefs have been suffering from, such as warming water (which causes another threat to reefs known as coral bleaching), pollution, and overfishing.

Oysters, Mussels, Urchins and Starfish

Generally, shelled animals—including mussels, clams, urchins and starfish—are going to have trouble building their shells in more acidic water, just like the corals. Mussels and oysters are expected to grow less shell by 25 percent and 10 percent respectively by the end of the century. Urchins and starfish aren't as well studied, but they build their shell-like parts from high-magnesium calcite, a type of calcium carbonate that dissolves even more quickly than the aragonite form of calcium carbonate that corals use. This means a weaker shell for these organisms, increasing the chance of being crushed or eaten.

Ochre seastars (Pisaster ochraceus) feed
on mussels off the coast of Oregon.

Some of the major impacts on these organisms go beyond adult shell-building, however. Mussels' byssal threads, with which they famously cling to rocks in the pounding surf, can't hold on as well in acidic water. Meanwhile, oyster larvae fail to even begin growing their shells. In their first 48 hours of life, oyster larvae undergo a massive growth spurt, building their shells quickly so they can start feeding. But the more acidic seawater eats away at their shells before they can form; this has already caused massive oyster die-offs in the U.S. Pacific Northwest.

This massive failure isn't universal, however: studies have found that crustaceans (such as lobsters, crabs, and shrimp) grow even stronger shells under higher acidity. This may be because their shells are constructed differently. Additionally, some species may have already adapted to higher acidity or have the ability to do so, such as purple sea urchins. (Although a new study found that larval urchins have trouble digesting their food under raised acidity.) Of course, the loss of these organisms would have much larger effects in the food chain, as they are food and habitat for many other animals.

Zooplankton

There are two major types of zooplankton (tiny drifting animals) that build shells made of calcium carbonate: foraminifera and pteropods. They may be small, but they are big players in the food webs of the ocean, as almost all larger life eats zooplankton or other animals that eat zooplankton. They are also critical to the carbon cycle—how carbon (as carbon dioxide and calcium carbonate) moves between air, land and sea. Oceans contain the greatest amount of actively cycled carbon in the world and are also very important in storing carbon. When shelled zooplankton (as well as shelled phytoplankton) die and sink to the seafloor, they carry their calcium carbonate shells with them, which are deposited as rock or sediment and stored for the foreseeable future. This is an important way that carbon dioxide is removed from the atmosphere, slowing the rise in temperature caused by the greenhouse effect.

This pair of sea butterflies (Limacina helicina) flutter
not far from the ocean's surface in the Arctic.

These tiny organisms reproduce so quickly that they may be able to adapt to acidity better than large, slow-reproducing animals. However, experiments in the lab and at carbon dioxide seeps (where pH is naturally low) have found that foraminifera do not handle higher acidity very well, as their shells dissolve rapidly. One study even predicts that foraminifera from tropical areas will be extinct by the end of the century.

The shells of pteropods are already dissolving in the Southern Ocean, where more acidic water from the deep sea rises to the surface, hastening the effects of acidification caused by human-derived carbon dioxide. Like corals, these sea snails are particularly susceptible because their shells are made of aragonite, a delicate form of calcium carbonate that is 50 percent more soluble in seawater.

One big unknown is whether acidification will affect jellyfish populations. In this case, the fear is that they will survive unharmed. Jellyfish compete with fish and other predators for food—mainly smaller zooplankton—and they also eat young fish themselves. If jellyfish thrive under warm and more acidic conditions while most other organisms suffer, it's possible that jellies will dominate some ecosystems (a problem already seen in parts of the ocean).

Plants and Algae

Neptune grass (Posidonia oceanica) is a slow-growing
and long-lived seagrass native to the Mediterranean.

Plants and many algae may thrive under acidic conditions. These organisms make their energy from combining sunlight and carbon dioxide—so more carbon dioxide in the water doesn't hurt them, but helps.

Seagrasses form shallow-water ecosystems along coasts that serve as nurseries for many larger fish, and can be home to thousands of different organisms. Under more acidic lab conditions, they were able to reproduce better, grow taller, and grow deeper roots—all good things. However, they are in decline for a number of other reasons—especially pollution flowing into coastal seawater—and it's unlikely that this boost from acidification will compensate entirely for losses caused by these other stresses.

Some species of algae grow better under more acidic conditions with the boost in carbon dioxide. But coralline algae, which build calcium carbonate skeletons and help cement coral reefs, do not fare so well. Most coralline algae species build shells from the high-magnesium calcite form of calcium carbonate, which is more soluble than the aragonite or regular calcite forms. One study found that, in acidifying conditions, coralline algae covered 92 percent less area, making space for other types of non-calcifying algae, which can smother and damage coral reefs. This is doubly bad because many coral larvae prefer to settle onto coralline algae when they are ready to leave the plankton stage and start life on a coral reef.

One major group of phytoplankton (single celled algae that float and grow in surface waters), the coccolithophores, grows shells. Early studies found that, like other shelled animals, their shells weakened, making them susceptible to damage. But a longer-term study let a common coccolithophore (Emiliania huxleyi) reproduce for 700 generations, taking about 12 full months, in the warmer and more acidic conditions expected to become reality in 100 years. The population was able to adapt, growing strong shells. It could be that they just needed more time to adapt, or that adaptation varies species by species or even population by population.

Fish

Two bright orange anemonefish poke their
heads between anemone tentacles.

While fish don't have shells, they will still feel the effects of acidification. Because the surrounding water has a lower pH, a fish's cells often come into balance with the seawater by taking in carbonic acid. These changes the pH of the fish's blood, a condition called acidosis.

Although the fish is then in harmony with its environment, many of the chemical reactions that take place in its body can be altered. Just a small change in pH can make a huge difference in survival. In humans, for instance, a drop in blood pH of 0.2-0.3 can cause seizures, comas, and even death. Likewise, a fish is also sensitive to pH and has to put its body into overdrive to bring its chemistry back to normal. To do so, it will burn extra energy to excrete the excess acid out of its blood through its gills, kidneys and intestines. It might not seem like this would use a lot of energy, but even a slight increase reduces the energy a fish has to take care of other tasks, such as digesting food, swimming rapidly to escape predators or catch food, and reproducing. It can also slow fish's growth.

Even slightly more acidic water may also affects fishes' minds. While clownfish can normally hear and avoid noisy predators, in more acidic water, they do not flee threatening noise. Clownfish also stray farther from home and have trouble "smelling" their way back. This may happen because acidification, which changes the pH of a fish's body and brain, could alter how the brain processes information. Additionally, cobia (a kind of popular game fish) grow larger otoliths—small ear bones that affect hearing and balance—in more acidic water, which could affect their ability to navigate and avoid prey.

Geological Oceanography

Geological oceanography is the study of Earth beneath the oceans. It involves geochemical, geophysical, sedimentological and paleontological investigations of the ocean floor and coastal margins. Geological oceanography includes exploring the ocean floor and the processes that form its canyons, valleys and mountains. Geological oceanographers research on the sea-floor spreading, plate tectonics, and oceanic circulation and climates. They examine the various ocean features such as rises and ridges, seamounts, trenches, etc. Geological oceanography is one of the broadest fields in the Earth Sciences and contains many subdisciplines, including geophysics and plate tectonics, petrology and sedimentation processes, and micropaleontology and stratigraphy.

The surface area of our planet Earth is covered by approximately 70% water that perhaps would have been named as a super water continent Oceania. The total area of the oceans and adjacent seas, forming a series of interconnected saline bodies is over 375.55×10^6 km^2. The oceans and seas are not evenly-distributed in the Northern and Southern Hemispheres: approximately 61% of the surface of the Northern Hemisphere is covered by ocean waters; about 81% of the Southern Hemisphere is covered. If calculated by volume, the total water of the oceans and seas is nearly 1.5×10^9 km^3. Our planet Earth is the only known body in the Solar System that is surrounded by waters filled with unique geological structures. The average depth of the ocean waters is approximately 4 km, but the depths of the oceans vary over a very wide range. In contrast to our general thinking, the oceans are not the deepest in their middle portions, but rather the great depths occur in trenches found along the continental margins. The greater depths of the oceans are found in trenches at the margins of the oceans. Excepting the Indonesia, Antilles, and Scotia deeps, major trenches are found in the Pacific Ocean floor. The greatest known depth is the Mariana Trench of the western Pacific Ocean, which is over 11km below sea level. The oceans are shallow near the sea level on continental shelves (~200 m). The continental slopes and rises are of intermediate depths (1 - 5 km) which are connecting the shelves and the deep sea, the abyssal plain, with average depths 4 km. The deep-sea morphology of the abyssal plain is characterized and shaped by channels, seamounts, mid-ocean ridges, and rift valleys, which are important components involved in the dynamics of the Earth.

The morphology of the sea floor features a rugged landscape which is not parallel to anywhere on the continents. Vast deep-sea mountains such as the mid-ocean ridge range are much more extensive than those on the continent. The mid-ocean ridge is the most prominent feature on the planet, occupying over an area larger than that covered by all continental mountains combined. The sea floor is continuously being created at the mid-ocean ridge, where newly-generated rocks of the oceanic crust melted out the mantle, and being consumed in the deep-sea trenches along the continental margins.

The processes of ocean crust generation/consumption, called sea floor spreading/subduction, play fundamental roles in global tectonics and account for ultimate geological forces that continuously shape the planet. The concept of the dynamic processes that shape the sea floor, being called plate tectonics, is now used for explaining most of the important morphological features of ocean basins. The observation of collecting data for supporting the concept of plate tectonics is mainly based on marine geological exploration, which was started in the 1970s. Marine geophysicists who generate, process, and interpret measurements of the Earth's physical properties, such as magnetic, gravity, seismic, electrical, electromagnetic, thermal, or radioactive patterns of the sea floor, have made great contributions in the studies aimed at predicting the dynamic nature of the Earth.

Most exploration geophysics is conducted to find commercial accumulations of oil, gas, coal, or other economical minerals. Oil and gas are finite, non-renewable resources. Most of the oil and gas are believed to be formed under special conditions by the decay of marine planktonic remains in sedimentary basins. It is hardly surprising that oil reservoirs should exist in the basins beneath the sea. The applications of marine geophysical techniques are by the reason, still important, as human population grows inexorably and the demand for energy rises. Advanced drilling technology has thus developed for exploring oil resources in basins of deeper waters.

Geological oceanographers are greatly benefited by the stride of systematic exploration of the sea floor by scientific drilling since 1970. Beginning from the first scientific cruise by GLOMAR Challenger by 1968, our knowledge on sediment distributions on the sea floor has greatly accumulated. The still on-going drilling program ODP (Ocean Drilling Program), evolving from many different phases of scientific drilling projects and using drilling vessel JOIDES Resolution, is the creation of an international community of marine geologists and geological oceanographers. By the drilling platform used in ODP, scientists are able to sample sediments and oceanic crusts to understand scientific problems such as how the ocean/atmosphere systems evolved since late Mesozoic; the complex interactive tectonic, magmatic, hydrothermal, and biological processes involved in the formation of new oceanic crust and lithosphere; and the geophysical and geochemical structures of oceanic crust and of more deeper lithosphere layers.

The Earth's environment, the fluid outer shell and land surface, appears to be changing rapidly and human activities are contributing to these changes. We now are gradually accept that the

Earth's system is high-dynamic and is characterized by multiple interactions among atmosphere, hydrosphere, lithosphere, and biosphere. The sediments accumulated on ocean basins provide us historical archives over geological time scales to understand the long-term behaviors of the dynamic geo-marine environment. Scientific drilling such as ODP enables us to better understand how the global environmental system operates and how sensitive the life-supporting system is to perturbations to its boundary conditions.

Marine Sedimentation

Marine sediments are the products of a limited number of physical, biological, and chemical processes. The nature of the resultant sediments is determined by the relative rates of input of material supplied by these processes.

Physical processes dominate at ocean margins, where they transfer particles eroded from the land to the sea floor. Active sedimentation processes (where the sediment modifies the properties and behavior of the suspension) include mass wasting and density currents. Such deposition tends to mask other sedimentary processes. The giant landslides surrounding the Hawaiian Islands and the abyssal plains of the Atlantic are striking deep-sea examples of active sedimentation. Because active sedimentation depends on gravitational energy, it does not extend seaward of the trenches along convergent plate boundaries (but can affect the entire ocean basin off passive margins).

Passive sedimentation processes are those in which the sediment is carried by but does not modify the normal thermohaline circulation. Examples:

- Hemipelagic sedimentation, where it appears that fine sediment moves along isopycnal surfaces high in the water column. This process is poorly understood (for example, the role of internal waves in resuspending sediment, the role of squirts and jets in transporting sediment offshore and the role of biota in sediment deposition have yet to be quantified). Hemipelagic sedimentation produces a fringe of terrigenous (land-derived) deposits up to a few hundred kilometers wide around land masses. These deposits are draped uniformly over the sea floor topography.

- "Drift" deposition along the path of bottom currents. Prominent in the high latitude North Atlantic along the flanks or ridges. Sediment in transit to drift deposits creates a near-bottom "nepheloid layer." Paleoceanography, has a dozen papers on drift deposits.

- Eolian (wind transported) sedimentation prominent where major wind systems cross semi-arid source areas (ephemeral lakes) or active ash-generating volcanoes. Prominent in the North Pacific, North Atlantic and Arabian Sea.

- Ice rafting, restricted to high latitudes, but can carry coarse particles into the subarctic gyres, far from land.

Sediments formed by physical processes have distinctive acoustic signatures of military interest. Hence they have been much studied during the past 50 years. "Passive" sediments record the history of deep currents, volcanism, aridity, wind trajectories, and iceberg abundances and trajectories.

Drift deposits, which can accumulate at hundreds of meters per million years, yield some of the highest resolution paleoceanographic records.

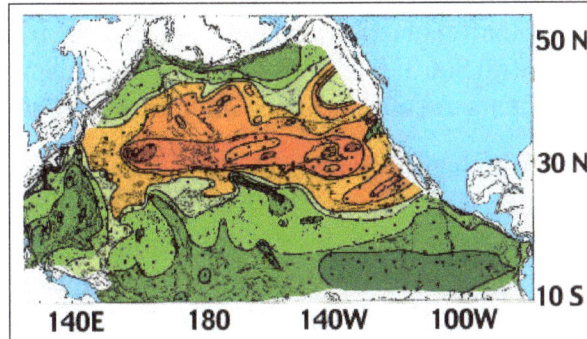

Quartz concentration in North Pacific sediments (carbonate-free basis)
showing effects of eolian transport in westerlies and northeast trades.

Biological processes dominate sediment formation in areas of high productivity that receives little terrigenous material. The equatorial Pacific and Southern Ocean are examples.

Several taxa of phytoplankton and zooplankton (the latter include benthic and upper water column representatives) secrete either $CaCO_3$ (coccolithophores and foraminiferans) or opal (hydrated SiO_2 - diatoms and radiolarians). If not masked by terrigenous material, the tests (shells) of these organisms can form carbonate or siliceous oozes.

Because the distribution and abundance of the various species are determined by the temperature, salinity, thermocline depth, carbonate chemistry, and productivity of the waters in which they live, most of what we know about paleoceanography is derived from fossil assemblages of these organisms.

In addition, the composition of the tests records the oxygen and carbon isotopic compositions and trace element contents of the waters in which they were secreted. These parameters in fossil tests provide insights to past ice volumes, temperature, productivity, and changes in biogeochemical cycles.

Benthic organisms also modify the historical record by actively mixing ("bioturbating") the most recently deposited sediments. Bioturbation, is effectively a low-pass filter that supresses or eliminates records of events that create layers of sediment thinner than the depth of mixing. Rapidly deposited or anoxic sediments provide the only deep-sea records capable of resolving events shorter than about a millennium.

Deep sea ferromanganese nodules on the floor of the South
Pacific Ocean (individual nodules are 5-10 cm diameter).

Chemical processes dominate sedimentation only in deep, low productivity areas shielded from terrigenous material. Precipitates from hydrothermal solutions emanating from mid-ocean ridges are prominent along the flanks of the East Pacific Rise in the South Pacific. Authigenic deposits (formed by the very slow precipitation of oxyhydroxides and silicates from normal seawater) form distinctive sediments in the central South Pacific as well as ferromanganese nodules and crusts on any surfaces where other sediments are absent or accumulating extremely slowly.

Chemical processes also modify biological (biogenic) sediments through the dissolution of $CaCO_3$ and (to a lesser extent) opal in deep water. T and P both play a role in increasing the corrosiveness of deep waters, but the major effect on carbonate tests is due to the creation of CO_2 by the biologically mediated oxidation of organic matter, which creates biocarbonate ions at the expense of carbonate ions (below equation), thereby driving the dissolution of carbonate tests.

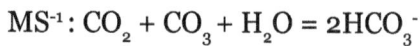

$$MS^{-1}: CO_2 + CO_3 + H_2O = 2HCO_3^-$$

In general, the carbonate content of deep sea sediments decreases with increasing water depth. Two important horizons are the lysocline, where the proportion of solution-resistant tests increases abruptly, and the calcite compensation depth (CCD) which is the boundary between carbonate-bearing and carbonate-free sediments.

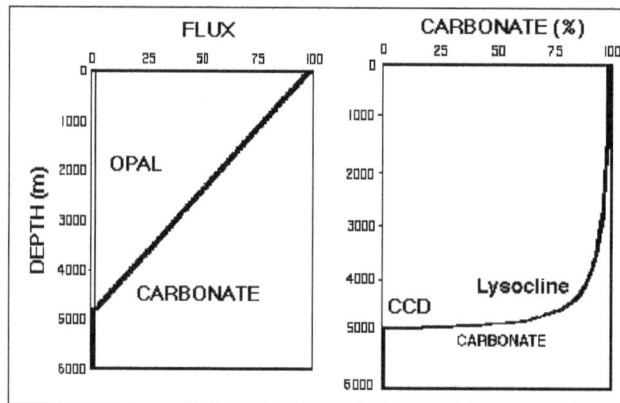

Carbonate concentration versus depth showing lysocline and CCD (right) created by simple linear dissolution with depth of carbonate in a carbonate-opal mixture (left).

The sea floor distributions of both carbonate and opal tests reflect the evolution of bottom water chemistry as it makes the "grand tour" from the North Atlantic via the Southern Ocean to the northern Pacific and Indian Oceans. As the bottom water "ages" it accumulates carbon dioxide, which drives equation MS^{-1} to the right, thereby dissolving $CaCO_3$.

Chemical reactions within the sea floor (diagenesis) can further modify deep-sea sediments, both through isochemical changes (which can obliterate particle source and paleomagnetic information) and open-system changes (which can modify the isotopic and trace element composition of sedimentary particles).

Rates of deposition- Sediments can be dated by a number of well-established techniques. The most common are:

   • First and last appearance of fossils.

- Paleomagnetic correlations.

- Decay of radioactive isotopes.

- Oxygen isotope correlations.

Physical and Chemical Properties of Sediment

Fall Velocity

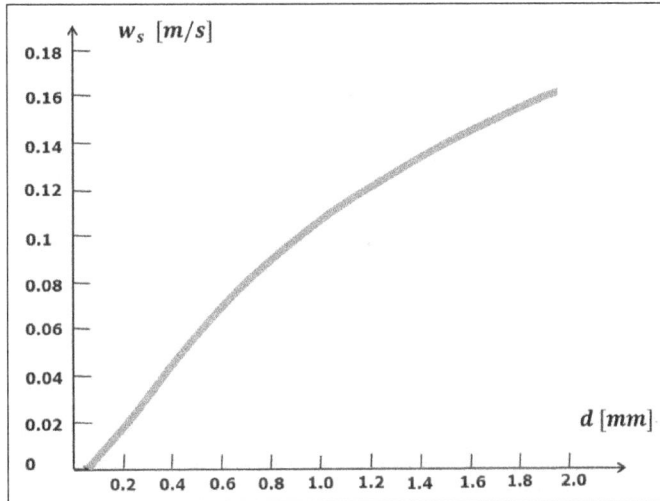

Fall velocity of quartz spheres in still water.

The seabed of oceans and coastal waters are mainly formed by settling of sediment particles out of suspension. Which particles settle where is mainly determined by the particle fall velocity. The fall velocity of sediment particles depends on size and density. Because the density of different types of sediments is similar, the grain size is determinant. The larger the grain size, the higher the fall velocity. Indicated is the fall velocity of round quartz grains in still water at low concentration. In reality, sediment particles do not have a round shape, which slightly affects the fall velocity. The actual fall velocity in flowing water is very different, due to up and down motions caused by turbulent eddies. The concentration of sediment particles also plays a role. At very high concentration, the fall velocity decreases, because particles increasingly interfere with each other, a phenomenon that is called hindered settling.

Flocculation

If the fall velocity for very small particles, such as clay minerals, is derived from settling quartz spheres, it seems so low that they almost never reach the bottom from suspension. The ubiquitous mud beds in coastal waters point to a different settling mechanism. This mechanism consists of flocculation. Clay minerals have a large surface area relative to their weight, and electrical surface charges. This ensures that they easily bind to one another and to other substances in the water. Clays are therefore classified as cohesive sediments. A substance that serves as a powerful binder is so-called EPS, extracellular polymeric substances. These large organic molecules (polysaccharides, proteins, nucleic acids and lipids) are exuded by living organisms and therefore omnipresent in coastal waters. In addition to EPS, bacterial colonization can also play a role in flocculation.

Flocculation is further influenced by factors such as salinity and pH of the water. When flocs grow, they not only capture clay particles, but also other suspended sediments, such as silt and fine sand. Frequent encounters between sediment particles are important for floc growth. Flocculation is thus enhanced with a high concentration of suspended material and with a certain degree of turbulence. However, when turbulence is too strong, flocs break off again. Flocs settle much faster than the individual constituent particles - a factor of a thousand or more, The largest flocs are agglomerates of microfloc. These so-called macroflocs have the highest fall velocity, but they are less stable than the smaller microflocs. The various factors that play a role are captured in the following empirical formulas, for macroflocs and microflocs respectively:

$$W_M = \frac{gB_M}{G}\left(\frac{c}{\rho}\right)^k\left(\frac{Gd_\mu^2}{v}\right)^{0.33}\exp\left[-\left(-\frac{u_{*M}}{\sqrt{\tau/\rho}}\right)^{0.463}\right]$$

$$W_\mu = \frac{gB_\mu}{G}\left(\frac{Gd^2}{v}\right)^{0.78}\exp\left[-\left(\frac{u\mu}{\sqrt{\tau/\rho}}\right)^{0.66}\right],$$

where, the index M designates the macroflocs and the index μ the microflocs.

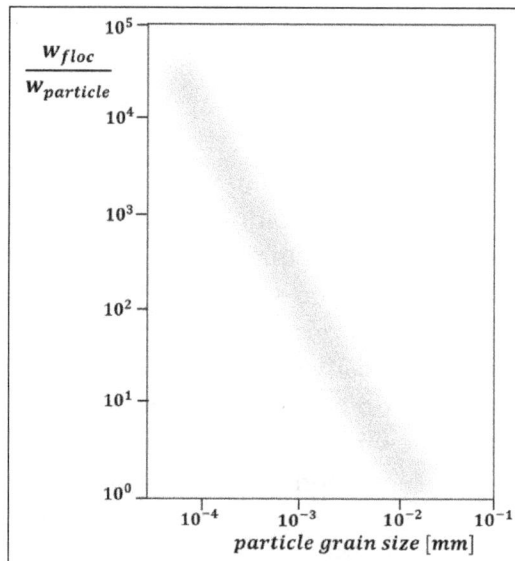

The ratio of floc fall velocity and fall velocity of the constituent particles.

Other symbols stand for: d the grainsize of the constituent particles, d_μ the grainsize of the constituent microflocs, τ is the bed shear stress and c the suspension concentration in kg/l. The velocity shear rate G is given by $G = \sqrt{\varepsilon/v} = \sqrt{\frac{\tau}{pv}}/\frac{dU}{dz}$.

Where, ε is the energy dissipation rate per unit mass and v the kinematic viscosity.

For the Tamar and Gironde estuaries the following parameter values were established: B_M = 0.13, B_μ = 0.6, k = 0.22, u_{*M} = 0.067m/s, $u_{*\mu}$ = 0.025m/s, d_μ = 10^{-4}m, d = 10^{-5} m. The settling velocities observed in the Tamar and Gironde are 0.5-1 mm/s for microflocs and about 5 times larger for macroflocs.

Sediment Deposits in Coastal Waters

Deposition of sediment on the seabed takes place when conditions are suitable. These conditions are different for each type of sediment.

When a sediment-laden water mass reaches an area of lower flow strength and wave activity, part of the carried material is deposited. Sediment particles with the greatest fall velocity settle first and particles entrained as bedload (rolling and jumping along the bottom) come to rest. When the current strength and wave activity further decrease, the fine suspended material also settles. Most sedimentation takes place in the period around slack tide (flow reversal). Part of the deposited material will afterwards be resuspended by the recovering tidal flow, but another part will remain. This can lead to temporary or permanent deposition. Deposits are temporary if they are insufficiently consolidated and re-eroded during conditions of strong currents (e.g., spring tide) or strong wave action (storm). Permanent deposits have a layered character; each layer represents a deposition period. Successive layers may contain different types of sediment if they have been deposited under different conditions.

Bedforms

Sediment layers are generally horizontal, but can also have a wavy character. These undulations are caused by the fact that the interaction between flow and sediment bed does not behave linearly, meaning that small disturbances of the flat sediment bed can grow exponentially. This leads to the emergence of a large range of bedforms, the smallest bedforms, ripples, arise in places where bed sediments are sandy and where currents and wave activity are not very strong but sufficient to set sediment in motion. For situations where sediment movement is mainly determined by waves. Bed ripples exert friction on tidal currents and have a great influence on the flow velocity. In addition to ripples, much larger bedforms can also arise: megaripples and dunes (also called sandwaves) with wavelengths ranging from several meters to more than hundred meters. In underlying sediment layers, bed ripples are usually not preserved, but megaripples and dunes can generally be observed quite well.

Bedforms do not occur with deposits of fine cohesive material (mud). Dewatering of freshly deposited mud is a slow process and mud layers therefore remain fluid for a long time as so-called fluid mud, Once consolidated, the erosion resistance is high, so that no bed ripples can form. Mud layers therefore have a smooth surface and exert little friction on the flow.

Graded Sediment

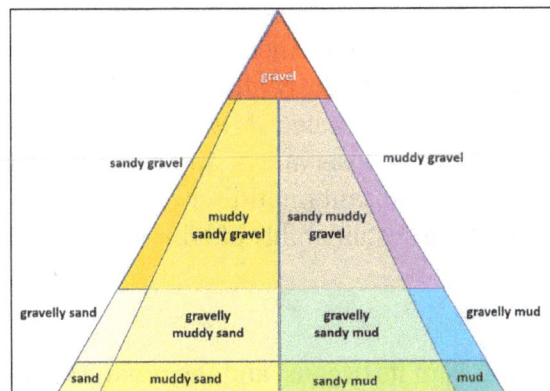

Classification of different types of graded sediment beds.

Sediment Transport

The processes that underlie sediment transport are strongly related to turbulent flow structures in a thin boundary layer near the bed and above. Because these processes are very complicated and not even fully understood, empirical formulas are used in practice for the description of sediment transport.

Sediment Contamination

As indicated earlier, fine sediments - clay particles in particular - can easily bind with other substances in the water. This binding is called sorption: adsorption (surface binding) or absorption (uptake). This property allows fine sediments to filter out dissolved contaminants from the water. The water is less polluted, but the contamination of bed sediments, on the other hand, increases. The partition of pollutants between the dissolved phase and the sediment-bound phase is represented by the parameter Kd. This partition, i.e. the value of Kd, depends on the type of pollutant and the type of sediment. Laboratory tests show that for inorganic contaminants, for example heavy metals such as lead (cation Pb^{2+}), cadmium (Cd^{2+}) and copper (Cu^{2+}), the partition depends on the sediment grainsize . The smaller the grainsize, the stronger is the sorption (large Kd), independently of the concentration of dissolved heavy metals. This does not apply to organic contaminants such as PCBs (polychlorinated biphenyls) and PAHs (polyciclic aromatic hydrocarbons), the sorption of which is determined by the organic carbon content in the sediment . This is a well-known feature in water sanitation technology, where active carbon is used to filter contaminants from the water.

Contaminants are much less toxic to marine organisms when they are bound to sediment than when they are dissolved in water . The bioavailability even decreases as the sediment bed gets older. Perturbation of the sediment bed by dredging or by bioturbation, however, plays a role. When sediment is worked up from a deeper anoxic layer to a higher oxic layer, attached metals are released by desorption. Burial from sediments to deeper soil layers, on the other hand, reduces bioavailability. The bioavailability of contaminants in soils can be reduced by adding active carbon.

Physical Oceanography

Physical oceanography focuses on describing and understanding the evolving patterns of ocean circulation and fluid motion, along with the distribution of its properties such as temperature, salinity and the concentration of dissolved chemical elements and gases. The ocean as a dynamic fluid is studied at a wide range of spatial scales, from the centimeter scales relevant to turbulent microstructure through the many thousand kilometer scales of the ocean gyres and global overturning circulation. Approaches include theory, direct observation, and computer simulation. Our research frequently takes place in the context of important multidisciplinary issues including the dynamics and predictability of global climate and the sustainability of human use in coastal and estuarine regions.

Physical Setting

Roughly 97% of the planet's water is in its oceans, and the oceans are the source of the vast majority of water vapor that condenses in the atmosphere and falls as rain or snow on the continents.

The tremendous heat capacity of the oceans moderates the planet's climate, and its absorption of various gases affects the composition of the atmosphere. The ocean's influence extends even to the composition of volcanic rocks through seafloor metamorphism, as well as to that of volcanic gases and magmas created at subduction zones.

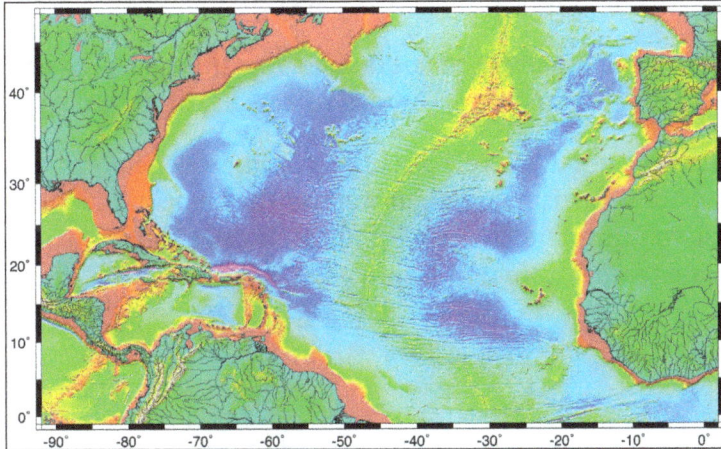

Perspective view of the sea floor of the Atlantic Ocean and the Caribbean Sea. The purple sea floor at the center of the view is the Puerto Rico Trench.

The oceans are far deeper than the continents are tall; examination of the Earth's hypsographic curve shows that the average elevation of Earth's landmasses is only 840 metres (2,760 ft), while the ocean's average depth is 3,800 metres (12,500 ft). Though this apparent discrepancy is great, for both land and sea, the respective extremes such as mountains and trenches are rare.

Table: Area, volume plus mean and maximum depths of oceans (excluding adjacent seas).

Body	Area (10^6km^2)	Volume (10^6km^3)	Mean depth (m)	Maximum (m)
Pacific Ocean	165.2	707.6	4282	-11033
Atlantic Ocean	82.4	323.6	3926	-8605
Indian Ocean	73.4	291.0	3963	-8047
Southern Ocean	20.3			-7235
Arctic Ocean	14.1		1038	
Caribbean Sea	2.8			-7686

Temperature, Salinity and Density

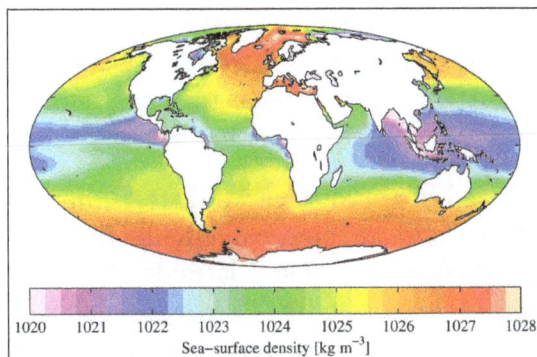

WOA surface density.

Because the vast majority of the world ocean's volume is deep water, the mean temperature of seawater is low; roughly 75% of the ocean's volume has a temperature from 0° – 5 °C. The same percentage falls in a salinity range between 34–35 ppt (3.4–3.5%). There is still quite a bit of variation, however. Surface temperatures can range from below freezing near the poles to 35 °C in restricted tropical seas, while salinity can vary from 10 to 41 ppt (1.0–4.1%).

The vertical structure of the temperature can be divided into three basic layers, a surface mixed layer, where gradients are low, a thermocline where gradients are high, and a poorly stratified abyss.

In terms of temperature, the ocean's layers are highly latitude-dependent; the thermocline is pronounced in the tropics, but nonexistent in polar waters. The halocline usually lies near the surface, where evaporation raises salinity in the tropics, or meltwater dilutes it in polar regions. These variations of salinity and temperature with depth change the density of the seawater, creating the pycnocline.

Circulation

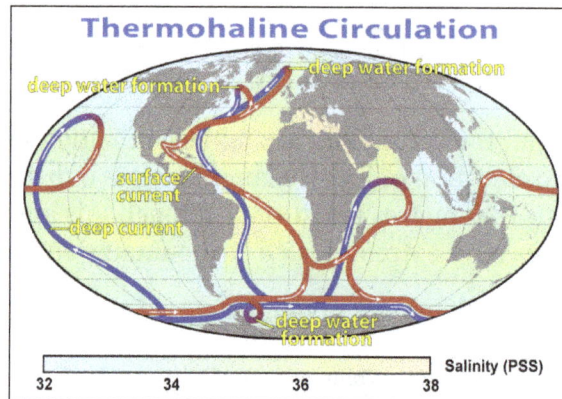

Density-driven thermohaline circulation.

Energy for the ocean circulation (and for the atmospheric circulation) comes from solar radiation and gravitational energy from the sun and moon. The amount of sunlight absorbed at the surface varies strongly with latitude, being greater at the equator than at the poles, and this engenders fluid motion in both the atmosphere and ocean that acts to redistribute heat from the equator towards the poles, thereby reducing the temperature gradients that would exist in the absence of fluid motion. Perhaps three quarters of this heat is carried in the atmosphere; the rest is carried in the ocean.

The atmosphere is heated from below, which leads to convection, the largest expression of which is the Hadley circulation. By contrast the ocean is heated from above, which tends to suppress convection. Instead ocean deep water is formed in polar regions where cold salty waters sink in fairly restricted areas. This is the beginning of the thermohaline circulation.

Oceanic currents are largely driven by the surface wind stress; hence the large-scale atmospheric circulation is important to understanding the ocean circulation. The Hadley circulation leads to Easterly winds in the tropics and Westerlies in mid-latitudes. This leads to slow equatorward flow throughout most of a subtropical ocean basin (the Sverdrup balance). The return flow occurs in an intense, narrow, poleward western boundary current. Like the atmosphere, the ocean is far wider than it is deep, and hence horizontal motion is in general much faster than vertical motion. In the southern hemisphere there is a continuous belt of ocean, and hence the mid-latitude westerlies force the strong Antarctic Circumpolar Current. In the northern hemisphere the land masses prevent this and the ocean circulation is broken into smaller gyres in the Atlantic and Pacific basins.

Coriolis Effect

The Coriolis effect results in a deflection of fluid flows (to the right in the Northern Hemisphere and left in the Southern Hemisphere). This has profound effects on the flow of the oceans. In particular it means the flow goes around high and low pressure systems, permitting them to persist for long periods of time. As a result, tiny variations in pressure can produce measurable currents. A slope of one part in one million in sea surface height, for example, will result in a current of 10 cm/s at mid-latitudes. The fact that the Coriolis effect is largest at the poles and weak at the equator results in sharp, relatively steady western boundary currents which are absent on eastern boundaries.

Ekman Transport

Ekman transport results in the net transport of surface water 90 degrees to the right of the wind in the Northern Hemisphere, and 90 degrees to the left of the wind in the Southern Hemisphere. As the wind blows across the surface of the ocean, it "grabs" onto a thin layer of the surface water. In turn, that thin sheet of water transfers motion energy to the thin layer of water under it, and so on. However, because of the Coriolis Effect, the direction of travel of the layers of water slowly move farther and farther to the right as they get deeper in the Northern Hemisphere, and to the left in the Southern Hemisphere. In most cases, the very bottom layer of water affected by the wind is at a depth of 100 m – 150 m and is traveling about 180 degrees, completely opposite of the direction that the wind is blowing. Overall, the net transport of water would be 90 degrees from the original direction of the wind.

Langmuir Circulation

Langmuir circulation results in the occurrence of thin, visible stripes, called windrows on the surface of the ocean parallel to the direction that the wind is blowing. If the wind is blowing with more than 3 ms^{-1}, it can create parallel windrows alternating upwelling and down welling about 5–300 m apart. These windrows are created by adjacent ovular water cells (extending to about 6 m (20 ft) deep) alternating rotating clockwise and counter clockwise. In the convergence zones debris, foam and seaweed accumulates, while at the divergence zones plankton are caught and carried to the surface. If there are many plankton in the divergence zone fish are often attracted to feed on them.

Ocean–atmosphere Interface

At the ocean-atmosphere interface, the ocean and atmosphere exchange fluxes of heat, moisture and momentum.

Heat

The important heat terms at the surface are the sensible heat flux, the latent heat flux, the incoming solar radiation and the balance of long-wave (infrared) radiation. In general, the tropical oceans will tend to show a net gain of heat, and the polar oceans a net loss, the result of a net transfer of energy polewards in the oceans.

Hurricane Isabel east of the Bahamas.

The oceans' large heat capacity moderates the climate of areas adjacent to the oceans, leading to a maritime climate at such locations. This can be a result of heat storage in summer and release in winter; or of transport of heat from warmer locations: a particularly notable example of this is Western Europe, which is heated at least in part by the north atlantic drift.

Momentum

Surface winds tend to be of order meters per second; ocean currents of order centimeters per second. Hence from the point of view of the atmosphere, the ocean can be considered effectively stationary; from the point of view of the ocean, the atmosphere imposes a significant wind stress on its surface, and these forces large-scale currents in the ocean.

Through the wind stress, the wind generates ocean surface waves; the longer waves have a phase velocity tending towards the wind speed. Momentum of the surface winds is transferred into the energy flux by the ocean surface waves. The increased roughness of the ocean surface, by the presence of the waves, changes the wind near the surface.

Moisture

The ocean can gain moisture from rainfall, or lose it through evaporation. Evaporative loss leaves the ocean saltier; the Mediterranean and Persian Gulf for example have strong evaporative loss; the resulting plume of dense salty water may be traced through the Straits of Gibraltar into the Atlantic Ocean. At one time, it was believed that evaporation/precipitation was a major driver of ocean currents; it is now known to be only a very minor factor.

Planetary Waves

Kelvin Waves

A Kelvin wave is any progressive wave that is channelled between two boundaries or opposing forces (usually between the Coriolis force and a coastline or the equator). There are two types, coastal and equatorial. Kelvin waves are gravity driven and non-dispersive. This means that Kelvin waves can retain their shape and direction over long periods of time. They are usually created by a sudden shift in the wind, such as the change of the trade winds at the beginning of the El Niño-Southern Oscillation.

Coastal Kelvin waves follow shorelines and will always propagate in a counterclockwise direction in the Northern hemisphere (with the shoreline to the right of the direction of travel) and clockwise in the Southern hemisphere.

Equatorial Kelvin waves propagate to the east in the Northern and Southern hemispheres, using the equator as a guide. Kelvin waves are known to have very high speeds, typically around 2–3 meters per second. They have wavelengths of thousands of kilometers and amplitudes in the tens of meters.

Rossby Waves

Rossby waves, or planetary waves are huge, slow waves generated in the troposphere by temperature differences between the ocean and the continents. Their major restoring force is the change in Coriolis force with latitude. Their wave amplitudes are usually in the tens of meters and very large wavelengths. They are usually found at low or mid-latitudes.

There are two types of Rossby waves: barotropic and baroclinic. Barotropic Rossby waves have the highest speeds and do not vary vertically. Baroclinic Rossby waves are much slower.

The special identifying feature of Rossby waves is that the phase velocity of each individual wave always has a westward component, but the group velocity can be in any direction. Usually the shorter Rossby waves have an eastward group velocity and the longer ones have a westward group velocity.

Climate Variability

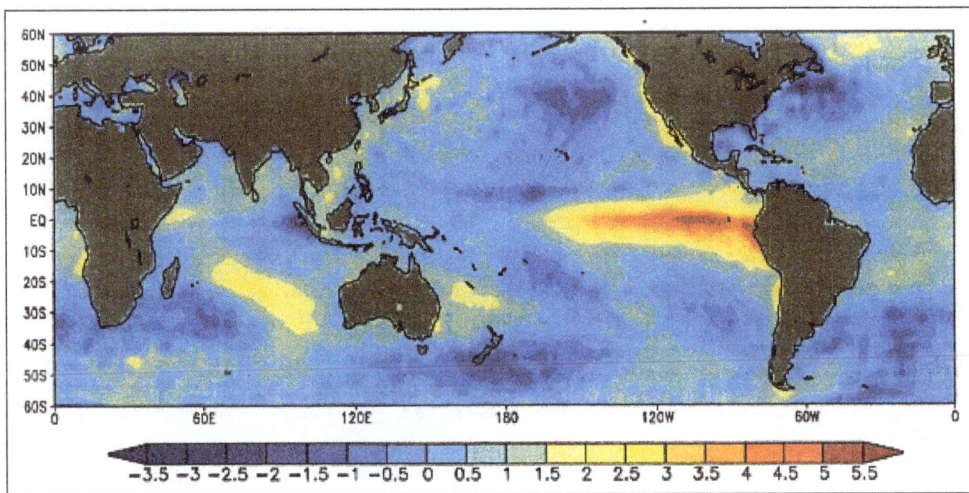

Ocean surface temperature anomaly [°C].

The interaction of ocean circulation, which serves as a type of heat pump, and biological effects such as the concentration of carbon dioxide can result in global climate changes on a time scale of decades. Known climate oscillations resulting from these interactions, include the Pacific decadal oscillation, North Atlantic oscillation, and Arctic oscillation. The oceanic process of thermohaline circulation is a significant component of heat redistribution across the globe, and changes in this circulation can have major impacts upon the climate.

Antarctic Circumpolar Wave

This is a coupled ocean/atmosphere wave that circles the Southern Ocean about every eight years. Since it is a wave-2 phenomenon (there are two peaks and two troughs in a latitude circle) at each fixed point in space a signal with a period of four years is seen. The wave moves eastward in the direction of the Antarctic Circumpolar Current.

Ocean Currents

Among the most important ocean currents are the:

- Antarctic Circumpolar Current

- Deep ocean (density-driven)

- Western boundary currents:

    ○ Gulf Stream

    ○ Kuroshio Current

    ○ Labrador Current

    ○ Oyashio Current

    ○ Agulhas Current

    ○ Brazil Current

    ○ East Australia Current

- Eastern Boundary currents:

    ○ California Current

    ○ Canary Current

    ○ Peru Current

    ○ Benguela Current

Antarctic Circumpolar

The ocean body surrounding the Antarctic is currently the only continuous body of water where there is a wide latitude band of open water. It interconnects the Atlantic, Pacific and Indian oceans, and provide an uninterrupted stretch for the prevailing westerly winds to significantly increase wave amplitudes. It is generally accepted that these prevailing winds are primarily responsible for the circumpolar current transport. This current is now thought to vary with time, possibly in an oscillatory manner.

Deep Ocean

In the Norwegian Sea evaporative cooling is predominant, and the sinking water mass, the North Atlantic Deep Water (NADW), fills the basin and spills southwards through crevasses in the submarine sills that connect Greenland, Iceland and Britain. It then flows along the western boundary of the Atlantic with some part of the flow moving eastward along the equator and then poleward into the ocean basins. The NADW is entrained into the Circumpolar Current, and can be traced into the Indian and Pacific basins. Flow from the Arctic Ocean Basin into the Pacific, however, is blocked by the narrow shallows of the Bering Strait.

Western Boundary

An idealised subtropical ocean basin forced by winds circling around a high pressure (anticyclonic) systems such as the Azores-Bermuda high develops a gyre circulation with slow steady flows towards the equator in the interior. As discussed by Henry Stommel, these flows are balanced in the region of the western boundary, where a thin fast polewards flow called a western boundary current develops. Flow in the real ocean is more complex, but the Gulf stream, Agulhas and Kuroshio are examples of such currents. They are narrow (approximately 100 km across) and fast (approximately 1.5 m/s).

Equatorwards western boundary currents occur in tropical and polar locations, e.g. the East Greenland and Labrador currents, in the Atlantic and the Oyashio. They are forced by winds circulation around low pressure (cyclonic).

Gulf Stream

The Gulf Stream, together with its northern extension, North Atlantic Current, is a powerful, warm, and swift Atlantic Ocean current that originates in the Gulf of Mexico, exits through the Strait of Florida, and follows the eastern coastlines of the United States and Newfoundland to the northeast before crossing the Atlantic Ocean.

Kuroshio

The Kuroshio Current is an ocean current found in the western Pacific Ocean off the east coast of Taiwan and flowing north-eastward past Japan, where it merges with the easterly drift of the North Pacific Current. It is analogous to the Gulf Stream in the Atlantic Ocean, transporting warm, tropical water northward towards the polar region.

Ocean Heat Content

The improper expression Oceanic heat content (OHC) refers to the heat absorbed by the ocean, which is then stored as a form of internal energy or enthalpy (because, in fact, heat is not a function of state and thus cannot be stored as it is in any way). Oceanography and climatology are the science branches which study ocean heat content. Changes in the ocean heat content play an important role in the sea level rise, because of thermal expansion. It is with high confidence that ocean warming accounts for 90% of the energy accumulation from global warming between 1971 and 2010. About one third of that extra heat has been estimated to propagate to depth below 700 meters. Beyond the direct impact of thermal expansion, ocean warming contributes to increased rates of ice melt of glaciers in fjords of Greenland and ice sheets in Antarctica. Warmer Oceans are also responsible for coral bleaching.

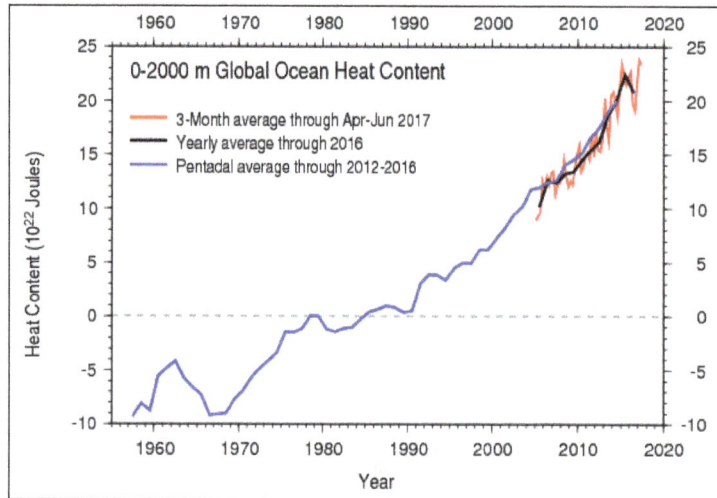

Global Heat Content (0–2000 meters) layer.

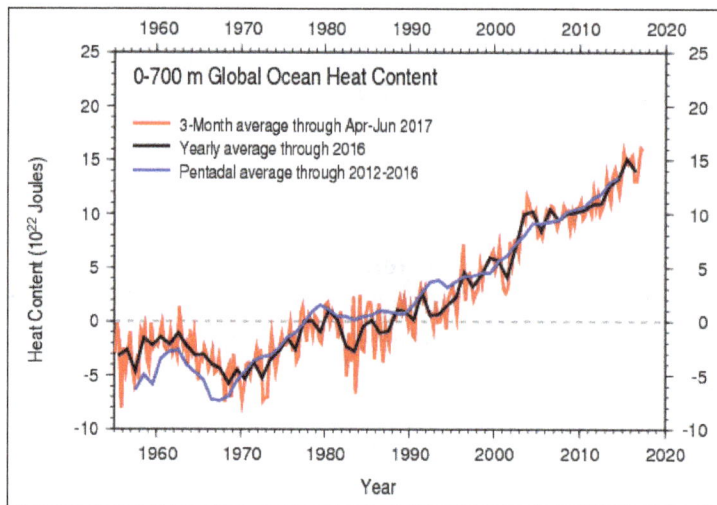

Global Heat Content (0–700 meters) layer.

Definition and Measurement

The areal density of ocean heat content between two depth levels is defined using a definite integral:

$$H = \rho c_p \int_{h2}^{h1} T(z)\,dz$$

Where, ρ is seawater density, c_p is the specific heat of sea water, $h2$ is the lower depth, $h1$ is the upper depth, and $T(z)$ is the temperature profile. In SI units, H has units of J·m⁻². Integrating this density over an ocean basin, or entire ocean, gives the total heat content, as indicated in the figure to right. Thus, the total heat content is the product of the density, specific heat capacity, and the volume integral of temperature over the three-dimensional region of the ocean in question.

Ocean heat content can be estimated using temperature measurements obtained by a Nansen bottle, an ARGO float, or ocean acoustic tomography. The World Ocean Database Project is the largest database for temperature profiles from all of the world's oceans.

The upper Ocean heat content in most North Atlantic regions is dominated by heat transport convergence (a location where ocean currents meet), without large changes to temperature and salinity relation.

Recent Changes

Several studies in recent years have found a multi-decadal oscillation increase in OHC of the deep and upper ocean regions and attribute the heat uptake to anthropogenic warming. Studies based on ARGO indicate that ocean surface winds, especially the subtropical trade winds in the Pacific Ocean, change ocean heat vertical distribution. This results in changes among ocean currents, and an increase of the subtropical overturning, which is also related to the El Niño and La Niña phenomenon. Depending on stochastic natural variability fluctuations, during La Niña years around 30% more heat from the upper ocean layer is transported into the deeper ocean. Model studies indicate that ocean currents transport more heat into deeper layers during La Niña years, following changes in wind circulation. Years with increased ocean heat uptake have been associated with negative phases of the interdecadal Pacific oscillation (IPO). This is of particular interest to climate scientists who use the data to estimate the ocean heat uptake.

Physical Oceanography of the Tropical Pacific

The tropics are a region of excess heating from the sun and towering cloud convection and rainfall in narrow bands across the Pacific. The warm surface waters of the tropics sustain the coral reefs which are a central feature of the islands' ecology. Compared with areas at higher latitudes, the frequency of storms and the average strength of the winds are low. As these affect the average height of surface waves, the wave climate of the tropics is relatively mild. Tsunamis (or "tidal waves") are a feature of life on a number of the Pacific islands which stand in the path of these fast-moving waves which are created by earthquakes or volcanoes.

The tropical Pacific is the seat of the global climate cycle known as "El Nino", which occurs every two to seven years. When the easterly trade winds weaken in the tropical Pacific, warm water builds up across the equatorial Pacific. These further changes the weather patterns in the atmosphere above, and the changes are propagated enormous distances around the globe through the atmosphere. The resulting disruption creates drought in some regions - the western tropical Pacific and northern Australia - and large rainfall in other regions - the eastern tropical Pacific and the west coasts of North and South America.

Surface Waves, Tides and Tsunamis

The ocean is constantly moving. Surface gravity waves are what catch our eyes - they are created by wind blowing over the sea surface either nearby (small or choppy waves) or far away (long ocean swell). We are also usually aware of the daily or twice daily cycle of tides, as beaches and reefs are successively covered and exposed. At times of the year when tides are very high and a storm creates large surface waves, "storm surges" can become a problem in low-lying coastal areas. Once in a long while, residents of coastal areas may be affected by a large and long-period wave called a "tsunami", generated by an earthquake either nearby or very far away.

Surface Waves

The wave climate of the Pacific Islands region is dominated by long period swell reaching the area from distant storms, by relatively low amplitude, short period waves generated by more local winds, and the occasional bursts of energy associated with intense local storms.

Waves are characterized by their wavelength (distance between crests or troughs), their period (time between successive passage of a crest past a fixed point), and their height or amplitude. Each type of wave can also be characterized by its restoring force. For surface waves, the restoring force to perturbations in sea surface height is gravity, and so the waves are sometimes referred to as surface gravity waves.

Surface waves are mostly created by wind blowing across the sea surface. The first waves to appear in response to wind are very small "capillary" waves, with wavelengths on the order of centimeters. These are apparent in a lake when a gust of wind blows past. If the wind persists, longer and longer waves are generated. The wave heights build, proportionally to the strength of the wind and how long it blows. Local waves forced by the wind travel in the direction of the wind. The period of a wave is the time between the passing of successive crests. For wind-generated waves, periods are on the order of seconds to many minutes for the shortest to the longest waves, respectively.

A large storm generates numerous surface waves moving in all different directions under the storm. These travel away from the storm location, so if the storm is localized, waves will radiate outwards from the storm area. The longer the waves, the faster they travel. Short waves are damped out much more rapidly by friction than are long waves. Long waves generated by storms at high latitudes, such as in the Gulf of Alaska, or far south in the Antarctic, or generated by earthquakes can travel clear across the Pacific without much attenuation.

Typically the sea state (field of waves) is a jumble of waves of many different wavelengths, moving in many directions since the wind forcing can be in many different directions. In the tropical Pacific, the wave field can be thought of as a superposition of waves forced by the local trade winds - the "tradewind sea" - and waves forced by distant storms. The tradewind sea is of small amplitude, and choppy since it is produced locally by winds which shift. The long period swell from far away storms is also of relatively low amplitude in the open ocean, and much more unidirectional than the tradewind sea.

The height of waves is now measured by various satellite sensors. A measure commonly used is "significant wave height", which is the average height of the highest one-third of the waves, where the height is measured from trough to crest. NASA routinely produces maps of significant wave height from satellite altimetry information. The altimeter measures the height of the sea surface, although the significant wave height is actually constructed from the properties of the radar pulse. Maps and information are available online, both for previous years and also in near real-time.

Monthly analyses for the globe show that the average wave height in the tropical Pacific is typically less than 3 meters, regardless of season, whereas wave height at high latitudes in the winter hemisphere typically reaches 3 to 6 meters due to large storms.

The water particles in a surface wave move in ellipses - up and forward in the direction of the wave propagation as the wave crest passes, and down and backwards as the trough passes. In deep water, waves with the longest wavelengths (distance from crest to crest) travel faster than short waves. When the wavelength becomes of the same size as the ocean bottom depth, the waves feel the bottom. The particle trajectories become more elliptical and the amplitude grows. The traveling speed of all waves becomes the same and proportional to the square root of the water depth - thus the waves travel more slowly and all together in shallower water.

As waves reach the shallow waters of a reef and island, they shoal, increase in amplitude and eventually break. The short period, tradewind sea produces relatively small surf height because of the short wavelengths. Large surf is produced by the long period swell from distant storms because of the correspondingly longer wavelength. The north shores of the Pacific islands receive this long-period swell in the northern hemisphere winter, and the south shores in the southern hemisphere winter. Wave heights of 6 meters in the surf zone are not uncommon. Winter swell on the north shore of Oahu occasionally reaches over 15 m.

Because the Pacific islands are small and rise steeply from the sea floor, there is little shelf area which can affect the progress of the long waves. (Continental shelves typically refract waves.) Thus the waves impinge directly on the shore or reef and do not wrap around the islands.

Breaking waves contain a lot of energy, some of which goes into production of local currents - first into longshore currents and then into rip currents which carry water back out to sea. Most of the circulation in the surf zone, and in lagoons inside reefs, is produced by breaking waves.

Tides

Tides are produced by the gravitational attraction between the earth and the moon and sun, and the centrifugal force on the earth as it moves around the center of gravity between it and the moon/sun. Since the orbits of these bodies are regular, tides are regular, and are in fact the only part of the ocean's motion which can be exactly predicted.

The complete tide is a composite of the moon (lunar) and sun (solar) tides. Considering just the moon, the gravitational between the earth and moon creates bulges of water on opposite sides of the earth. The water bulge nearest the moon/sun is due to domination of gravitational attraction over centrifugal force; the water bulge opposite the moon/sun is due to domination of centrifugal force. The two forces cancel at the earth's center. Since the earth also rotates daily, a point on the earth passes through these bulges twice a day, resulting in semi-diurnal (twice daily) components to the tide at each location. Because the moon and the sun do not generally lie over the equator, one of the bulges at a given point on the earth is larger than the other, leading to what is known as the "diurnal inequality", which lends a diurnal (daily) component to the tide.

A modulation of the tidal range results from the relative position of the moon and the sun: when the moon is new or full, the moon and the sun act together to produce larger "spring" tides; when the moon is in its first or last quarter, smaller "neap" tides occur. The cycle of spring to neap tides and back is half the 27-day period of the moon's revolution around the earth, and is known as the fortnightly cycle. The combination of diurnal, semi-diurnal and fortnightly cycles dominates variations in sea level throughout the islands.

The geometry of the oceans - the basin shape, local coastline, bays, and even harbor geometry - has a major effect on the local behavior of the tides. On scales of oceanic basins, tides exist as very long waves propagating in patterns determined by their period and the geometry of the basin. The response of the Pacific to the tidal period of 23 h 56 min, the largest diurnal (once daily) component. The tidal amplitude is very low in the central Pacific, but is higher in the tropical region of Australia, New Guinea and Indonesia, as well as far to the north in the Gulf of Alaska and subpolar region. Lines along which high tide occurs at the same time, converge to several points where the tidal range is zero. There are four of these points, called "amphidromes" in the Pacific: one on the North Pacific near the dateline, one near the equator in the eastern North Pacific, ond in the central South Pacific near Tahiti, and one east of New Zealand. Phase lines rotate counter-clockwise around the amphidromes in the North Pacific and clockwise around the ones in the South Pacific. For example, at the Hawaiian Islands, the offshore diurnal tide reaches the Hawai'i island first, then sweeps across Maui, O'ahu and finally Kauai.

Local bathymetry affects the ranges and phases of the tides along the shore, as the tidal waves wrap around the islands. For example, high tide at Haleiwa on the north shore of O'ahu occurs over an hour before high tide at Honolulu Harbor. Even though the tides at one point on a coastline are not in phase with those at even a nearby point, the tides at that point can be completely predicted if they are measured for several months, because the forcing which creates the tides is so extremely regular.

Tidal currents result from tidal variations of sea level, and near the shore is often stronger than the large scale circulation. Complete mapping of tidal currents requires direct measurements. As an example, the semidiurnal and diurnal tidal currents for Hawaii, show that the semi-diurnal and diurnal tidal currents tend to be aligned with the shoreline. Due to high variability of tidal currents around the islands, however, this statistical picture may not correspond to the flow at a particular time: tidal currents cannot be predicted as precisely as sea level. Strong swirls often result from tidal currents flowing around points and headlands, and present hazards to divers.

Tsunamis

When the seafloor is raised suddenly during a shallow earthquake, water is raised with it, producing a mound of excess water at the sea surface. Gravity collapses the mound, producing a series of waves: a tsunami. Tsunamis are gravity waves, just like those generated by the wind, but their period is much longer, on the order of 10 to 60 minutes. While earthquakes are the most common cause of tsunamis, the waves are generated by any phenomenon which rapidly changes the shape of the sea surface over a large area: volcanic eruption, landslide, even meteorite impact. Since the largest shallow earthquakes occur in the subduction zones which ring the Pacific, and since these same subduction zones are dotted with volcanoes, the tsunami hazard throughout the tropical Pacific is high.

On the open ocean, the wavelength of a tsunami may be as much as two hundred kilometers, many times greater than the ocean depth which is on the order of several kilometers. This huge wavelength means that the entire water column, from surface to bottom, is set into motion. Tsunamis therefore always behave like waves in shallow water, which means, their speed is proportional to the square root of the water depth. For typical ocean depths of 5 km, a tsunami will advance at 800 km/hr, about the speed of jet aircraft. A tsunami can therefore travel from one side of the Pacific to the other in less than a day. The speed decreases rapidly as the water shoals, in 15 m of water the speed of a tsunami (or of any wave with long enough wavelength to "feel" the ocean bottom) will be only 45 km/hr.

As the tsunami slows in shoaling water its wavelength is shortened. Just as with ordinary surf, the energy of the waves must be contained in a smaller volume of water, so the waves grow in height. The maximum height the tsunami reaches on shore is called the runup. Any runup over a meter is dangerous. Waves reaching only a meter above sea level may not seem threatening, but the waves of a tsunami are unlike normal waves. Even though the wavelength has shortened, a tsunami will typically have a wavelength in excess of ten kilometers when it comes ashore. Each wave therefore floods the land as a rapidly rising tide lasting for several minutes. The individual waves are typically from ten minutes to a half-hour apart, so the danger period can last for hours.

Runup can vary dramatically depending on seafloor topography. Small islands with steep slopes experience little runup; wave heights there are only slightly greater than on the open ocean. For this reason the smaller Polynesian islands with steep-sided fringing or barrier reefs are only at moderate hazard from tsunamis. Such is not the case for the Hawaiian Islands or the Marquesas, however. Both of these island chains are almost devoid of barrier reefs and have broad bays exposed to the open ocean. Hilo Bay at the island of Hawaii and Tahauku Bay at Hiva Oa are especially vulnerable. During a tsunami from the Eastern Aleutians in 1946, runup exceeded 8 m at Hilo and 10 m at Tahauku; 59 people were killed in Hilo, two in Tahauku. Similarly, any gap in a reef puts the adjacent shoreline at risk. The tsunami from the Suva earthquake of 1953 did little damage because of Fiji's extensive offshore reefs. Two villages on Viti Levu located opposite gaps in the reef, however, were extensively damaged and five people were drowned.

Tsunamis are generated by shallow earthquakes all around the Pacific, but those from earthquakes in the tropical Pacific tend to be modest in size. While such tsunamis may be devastating locally, they decay rapidly with distance and are usually not observed more than a few hundred kilometers from their sources. That is not the case with tsunamis generated by great earthquakes in the North Pacific or along the Pacific coast of South America. About half-a-dozen times a century a tsunami from one of these locations sweeps across the entire Pacific, is reflected from distant shores, and sets the entire ocean oscillating for days. The tsunami from the magnitude 9.5 Chile earthquake of 1960 caused death and destruction throughout the Pacific: Hawaii, Samoa, and Easter Island all recorded runups exceeding 4 m; 61 people were killed in Hawaii. In Japan 200 people died. A similar tsunami in 1868 from northern Chile caused extensive damage in the Austral Islands, Hawaii, Samoa, and New Zealand. There were several deaths in the Chatham Islands.

The tsunami from a local earthquake may reach a nearby shore in less than ten minutes, making warning a difficult task (though in this case the shaking of the ground provides its own warning). For tsunamis from more distant sources, however, accurate warnings of when a tsunami might arrive are possible because tsunamis travel at a known speed. The current international tsunami warning system has 26 member nations which coordinate their warning activities through the

Pacific Tsunami Warning Center in Hawaii. The Hawaii center uses seismic data from the global seismic network to identify and characterize potential tsunamigenic earthquakes, then verifies if a tsunami has been generated by querying tide gauge stations near the source. While the system is far from perfect (about half of the warnings are false alarms), performance is constantly improving and there have been no missed warnings.

Temperature and Salinity Distribution in the Tropical Pacific

The temperature of the sea has a large effect on local climate - what can grow in the water and on nearby land, fog and precipitation, production of hurricanes, and so on. The salt in seawater is what most obviously distinguishes it from freshwater, and affects the ecology of coastal lagoons, tidal flats, and river mouths. The salt has less overt influence than temperature on climate, but it does affect how deeply the surface layer of the ocean can mix and hence on the temperature of the surface layer, and thus has a subtle effect on climate.

Temperature

Ocean surface temperature globally is dominated by excess heating in the tropics compared with higher latitudes, resulting mainly from higher radiation from the sun in the tropics. This leads to a sea surface temperature difference from equator to pole of about 30 °C. In the tropics, including the tropical Pacific, the maximum sea surface temperature is around 28 °C and can rise to at most 30 °C. This is considerably cooler than the maximum temperatures regularly found over land, of about 50 °C. It is currently hypothesized that the main regulation on the maximum ocean temperature is through cloud formation. Cloud formation increases dramatically when the sea surface temperature is greater than about 27.5 °C. The increased cloudiness increases the albedo (reflectivity of the earth/atmosphere to space), which reduces the solar radiation reaching the sea surface, and thus keeps the surface temperature from rising much more.

The sea surface temperature is not uniformly high in the tropical Pacific. A large "warm pool" is found in the central and western Pacific, and also extends into the eastern Indian Ocean. Surface water in the eastern equatorial Pacific is several degrees cooler than in the west. The vertical thermal structure of the upper ocean is responsible for these differences. In the western Pacific, the surface layer, which is fairly well mixed, is approximately 100 meters thick, and warmer than about 28 °C. Just below this surface layer, the temperature changes rapidly downward; this is called the "thermocline". In the central and eastern Pacific, the surface layer is shallower, and so colder water and the thermocline are found closer to the surface. Upwelling in the eastern Pacific draws this cooler water to the surface, creating the equatorial "cold tongue" at the sea surface. Upwelling of cold water at the equator is apparent in sections crossing the equator. Upwelling in the western Pacific is somewhat weaker than in the east and draws up only warm water, and so an equivalent cold tongue along the equator is absent.

Upwelling is common along the west coast of South America, off Ecuador and Peru, and along the west coast of Central and North America. As a result of both the upwelling and the eastern boundary currents which flow towards the equator in these regions, sea surface temperatures are relatively low along these coasts. The winds which create upwelling are strongest in the area just west of Costa Rica. Here the thermocline is lifted to within 10 meters of the sea surface, and is called the Costa Rican Dome.

Below the sea surface, temperature decreases to the ocean bottom. The most rapid change is in the upper 500 meters, in the thermocline. Changes are more gradual below this. Temperature reaches about 1.2 °C in the abyssal tropical Pacific. The initial temperature and salinity of all ocean water is set at the sea surface. The sea surface temperature distribution shows that water colder than about 18 °C comes from latitudes higher than about 30°, hence outside the tropics. Waters of about 4-6 °C come from latitudes of about 40-45° (northern and southern hemisphere). The coldest waters flow northward from the Antarctic region. These southern hemisphere waters, which fill the Pacific below 1000 to 1500 meters, are part of a circulation which extends through all of the oceans. The deepest waters come from the the Weddell and Ross Seas of the Antarctic and the Greenland Sea just north of the North Atlantic. The North Pacific does not produce any of this deep water, and so its deep waters have traveled a long distance from their sea surface origin. These deep waters have spent about 500 years making the journey to the deep North Pacific (and slightly less time to the deep tropical Pacific). Waters which have been far from sea surface forcing (heating/cooling and evaporation/precipitation) for a long time are fairly uniform because they mix with each other. Thus the deep Pacific contains a large amount of water in a very narrow range of temperature and salinity, centered around 1.2 °C and 34.70 in practical salinity units. This water must upwell slowly and eventually complete the overturning cycle by reaching the sea surface, perhaps very far from the deep North Pacific.

Salinity

Sea water density depends on temperature (warm water is less dense), and also on the amount of material dissolved in the water. The latter is mostly what is referred to as "sea salt", and is a combination of various salts. The total amount of salt in the world ocean is constant on all but the longest geological timescales. However the total amount of fresh water in the ocean is not constant - it is affected by evaporation, precipitation and runoff. Hence salinity, which is more or less the grams of salts dissolved in a kilogram of seawater, varies as a result of surface freshwater inputs and exports.

The total range of salinity in most areas of the ocean is small enough that temperature actually contributes more to sea water density differences, but salinity differences are significant and important. For instance, if saltier water lies above fresher water, then the temperature difference between the two must be large enough to ensure stability (light water over dense water).

Surface salinity in the Pacific shows clearly the net result of the atmospheric circulation Cloud formation and high precipitation occur in regions of rising, humid air, which are associated with low atmospheric pressure at the sea surface, such as in the Intertropical Convergence Zone (ITCZ) at 5-10° N and subpolar regions poleward of 40°. Surface salinity is low where precipitation is high. Evaporation and hence surface salinity are high where the air is dry - regions of atmospheric divergence (high pressure zones at the surface).

Because temperature dominates the vertical density differences in the ocean, it decreases downward almost everywhere. Thus although salinity also contributes to density, the salinity distribution can be more complex, with regions of salty water lying over fresher water and vice versa. Such salinity inversions are common. In cross section from south to north, the high salinity in the surface evaporation cells extends down to the thermocline. The fresher water associated with the ITCZ extends fairly deep. Below the high salinity surface water is found a layer of low salinity "intermediate water" which extends from the rainy subpolar latitudes in the south and north towards the equator. Below this, the deep Pacific is filled with relatively more saline waters originating from the deep waters around Antarctica and from the Atlantic.

Along the equator surface salinity is lowest in the western Pacific, where normally there is much more rainfall than in the central and eastern equatorial Pacific. The freshest surface water in the western equatorial Pacific actually extends only partway down into the vertically-uniform, warm surface layer with salinity increasing strongly downward midway within this uniform temperature layer. Hence the surface stratification is dominated by salinity rather than temperature. A relatively sharp north-south front separates the fresh western equatorial surface water from the more saline central Pacific surface water. During periods such as El Nino when the trade winds slacken, the western fresh, warm water floods eastward towards the central Pacific along the equator.

Biological productivity in the ocean relies on nutrients in the sunlit surface layer (euphotic zone - about 100 meters depth). The principal nutrients which are routinely measured are nitrate, phosphate and dissolved silica. They are consumed by plants and animals in the ocean's surface layer. They are "regenerated" at depth as the decaying plants and animals and fecal pellets fall through the water column, with some portion, especially of the silica-bearing hard parts, reaching the ocean bottom. Thus nutrients are severely depleted in the surface layer where they are used almost as quickly as they appear there. Nutrients are found in abundance below the surface layer, especially where waters have been separated from the sea surface for a long time. Nutrients reach the euphotic zone through upwelling, and so upwelling regions have slightly higher nutrient content and much higher biological productivity than downwelling regions. The most productive regions occur where upwelling is vigorous and where the nutrient-rich thermocline is near the sea surface. Near-surface nutrients in the Pacific are high in the equatorial and eastern tropical Pacific where upwelling is high, and low in the subtropical downwelling regions poleward of about 20°. Surface nutrients are higher in the eastern equatorial Pacific than in the western, reflecting the upwelling of the thermocline waters towards the east.

Ocean Circulation in the Tropical Pacific

The ocean motions which are clear to a person on shore looking at the ocean. The ocean also has much slower motion - ocean currents which vary slowly over weeks to months, years and many decades. These affect navigation. Currents are also important in moving water from one place to another, which redistributes heat, salt, and higher and lower nutrients.

Surface Circulation

The Pacific sea surface circulation consists of two large "subtropical gyres" centered at 30°N and 30°S, which rotate clockwise in the northern hemisphere and counterclockwise in the southern hemisphere, a "subpolar gyre" centered at about 50°N and rotating counterclockwise, a major eastward flow which circles Antarctica called the "Antarctic Circumpolar Current", and complicated but predominantly zonal (east-west) currents in the tropics between about 15°N and 15°S. At the sea surface, flow is westward from 30°S up across the equator to about 5°N. This westward flow is all called the "South Equatorial Current". Between 5°N and 10°N lies a strong eastward flow, termed the North Equatorial Countercurrent. It is associated with and driven by the winds of the ITCZ. The westward flow between 10N and 30N is called the North Equatorial Current (NEC).

The northern half of the NEC is actually part of the subtropical gyre and the southern half is part of the ITCZ's elongated counterclockwise flow. Sometimes a weak ITCZ (South Pacific Convergence Zone) is also present in the southern hemisphere, creating an occasional appearance of a South Equatorial Countercurrent analogous to the North Equatorial Countercurrent.

In the western tropical Pacific, the circulation is dominated by strong currents which abut the western boundary. Western boundary currents are a central feature of all circulation patterns worldwide. In the tropical and South Pacific, the western boundary currents are complicated by the many islands and deep ridges. Australia forms the largest single part of the boundary. In the North Pacific, the westward-flowing North Equatorial Current reaches the western boundary at Mindanao in the Philippines. It splits into a northward flow, called the Kuroshio, and a southward flow, called the Mindanao Current. The Kuroshio flows into the East China Sea and then northward to the southern end of Japan (Kyushu) where it splits into a major flow eastward along the eastern coast of Japan, and a weaker flow, called the Tsushima Current, into the Japan East Sea. The Kuroshio is one of the strongest currents in the world, similar to the Gulf Stream and the Antarctic Circumpolar Current in strength. It affects climate in Japan through its warmth and fisheries off Japan through both its warmth and relative lack of nutrients. The Mindanao Current flows southward along Mindanao and separates to flow eastward into the North Equatorial Countercurrent at about 5°N. A portion turns westward at the southern end of Mindanao and enters the Celebes Sea.

"Eddies" (circulations of about 50 to 200 km size which are often variable over a period of weeks to months) are usually found east of Mindanao and east of Halmahera. The water entering the Celebes Sea forms the beginning of flow westward through the complex of Indonesian islands, threading through to Java and thence into the Indian Ocean.

In the South Pacific, the very broad, westward-flowing South Equatorial Current reaches the western boundary through a complex of islands. The northern portion forms a northward-flowing western boundary current along New Guinea, called the New Guinea Coastal Current. This flows northward to the equator. A portion of it turns eastward along the equator and apparently forms part of the eastward-flowing subsurface Equatorial Undercurrent. A portion may continue slightly northward, joined by the westward flow just north of the equator, and then turns eastward, joining the separated Mindanao Current, into the North Equatorial Countercurrent.

The remainder of the westward-flowing South Equatorial Current flows north of Fiji into the Coral Sea and reaches the western boundary at Australia. Here it turns southward into the East Australian Current, which is the western boundary current, and then flows southward to the northern tip of New Zealand. At this point, the current meanders a great deal and some portion of it separates and flows eastward just north of New Zealand as the North Cape Current. The broad flow between New Zealand and Fiji is also eastward.

The large-scale surface flow is affected only by the larger land masses, and not much by the small islands dotting the tropical and South Pacific. Intermediate and abyssal flow however are strongly affected by the ridges in which the small islands are embedded.

Subsurface Equatorial Circulation

The currents below the sea surface seem of less immediate importance to man, as they do not affect sailing or have an obvious effect on local ocean surface conditions such as temperature. However, the surface and deeper flows are strongly coupled to each other. It has become clear in recent years that successful computer models of the ocean circulation must include the flow below the surface, all the way down to the ocean bottom, where undersea rises and mountains strongly steer the bottom currents.

In most places of the world ocean, the currents vary only gradually from surface to bottom - they are usually strongest at the surface where they are closest to the wind forcing, and gradually blend into the circulation of the abyss. However, within 2 or 3° latitude of the equator, the subsurface currents are much more complicated. Between 100 and 200 meters depth lies the strong eastward-flowing Equatorial Undercurrent. The undercurrent was originally discovered by Townsend Cromwell during a research expedition in the 1950's when the drogues deployed at that depth moved strongly eastward while the surface current was westward. In speed, the Equatorial Undercurrent matches the strongest currents in the world (> 100 cm/sec or 1 km/day). However, the undercurrent is vertically very thin (about 100 meters thick) in contrast with the other major currents such as the Kuroshio, Gulf Stream, and Antarctic Circumpolar Current which reach to the ocean bottom.

The undercurrent and flanking it on either side of the equator lie the North and South Subsurface Countercurrents, flowing eastward (at 2° on either side of the equator and below 150 meters depth). These were discovered by Tsuchiya. Directly beneath the Equatorial Undercurrent lies a somewhat weaker westward flow, which extends to about 1000 meters depth. Below this there is a regime of the so-called "stacked jets", extending to the ocean bottom, but with vertical extent increasing towards the bottom. Farther away from the equator, between 2° and 5° latitude, the vertical structure may show only a reversal or two. Farther away from the equator than this, the vertical structure is much simpler, with the surface circulation extending to depths of 1000 to 2500 meters, and much weaker flow dominated by bottom topography.

The most general characteristic of circulation in the tropical Pacific is the exaggerated east-west nature compared with flow poleward of 20° latitude in both hemispheres, where "gyres" which also include more north-south flow are the norm. This zonality is characteristic of the tropical circulation in the Atlantic and Indian Oceans as well as the Pacific.

Deep Circulation

With increasing depth, the surface circulation weakens and shifts latitude. In the tropics, the surface circulation signatures disappear by about 500 to 1000 meters depth. Flow beneath this is predominantly zonal (east-west) with very slight north-south movement. Various analyses show counterclockwise circulation north of the equator and clockwise circulation south of the equator, in very elongated cells between the equator and about 10° latitude. The deepest circulation is affected by the topography of the ridges and basins. Overall, there is net northward flow in a deep western boundary current, which enters the Pacific from the Antarctic east of New Zealand and passes through a deep gap near Samoa, called the Samoan Passage. It moves on northward to the equator, crossing in the western Pacific. North of the equator, a portion branches eastward to pass south of the Hawaiian Islands, and the other portion continues northward. The northward flow appears to move westward under the Kuroshio and then northward along the western boundary to the subpolar Pacific. Return flow to the south probably occurs along the East Pacific Rise in the eastern Pacific and then westward along the equator.

Circulation Near Islands and Island Groups

Local circulation near islands and island chains can be affected by eddies generated by the ocean currents moving past the islands. Large island groups and especially the ridges upon which they sit also affect the large-scale ocean circulation. An example is flow near the Hawaiian Islands, which form a ridge for deep flow. On the north side of the Hawaiian Islands, large-scale currents or large eddies (time-dependent currents of possibly smaller spatial extent) are sometimes found along the ridge. An eddy is often generated at the passage between the islands of Maui and Hawaii. Southwest of the ridge, in the lee of the flow of ocean currents towards the west, eddy activity is reduced.

Forcing of the Circulation

All movement of ocean water must be generated by some force. Surface waves are created by the wind blowing over the sea surface and catching on smaller waves to make larger ones. Tides are created by the gravitational pull between the earth and the moon and sun. Tsunamis are created by undersea earthquakes. Ocean currents and large eddies are created by the winds acting much more indirectly than for surface waves, and also by cooling and evaporation which can cause the water to overturn.

The upper ocean circulation in the tropical Pacific is driven mostly by the stress from the wind. The prevailing winds in the tropical Pacific are the trades or easterlies, which blow from east to west. Together with the westerlies of higher latitudes, these force the large subtropical gyres. The dominant influence of these gyres on the tropics is the broad-scale westward flow, called the North Equatorial Current (north of 5°N) and the South Equatorial Current (from the equator southward). We divide the wind forcing of the tropics into two regimes - off the equator and on the equator. The difference between these is the importance of the earth's rotation to the forcing - off the equator it is very important and on the equator we can disregard it.

Wind Forcing of Non-equatorial Flow

The mechanism for forcing the large-scale circulation by the surface wind stress is indirect, and described well in introductory texts on physical oceanography. The large scale circulation is in "geostrophic balance". This means that the currents are driven by horizontal pressure differences which are balanced by the Coriolis force, which comes from the earth's rotation. The resulting flow is exactly at right angles to the pressure difference force- in the northern hemisphere it is to the right (so flow circulates around high pressure in a clockwise direction) and in the southern hemisphere it is to the left (counterclockwise flow around a high). Near the sea surface, the pressure difference is due to small, but large-scale and long-lasting, differences in sea surface height. Over about 100 kilometers horizontal distance, which is the width of a major current such as the Kuroshio, the sea surface height difference which creates the pressure difference which drives the current is no more than 1 meter. This is of course shorter than most surface waves in mid-ocean. The distinction between the surface height difference which drives a major current and that of just a surface wave is that the wave is just passing by - it changes the sea surface height over a very small time, whereas the surface height differences which drive currents must be in place for at least several days in order to "feel" the rotation of the earth. The largest height changes drive the fastest currents, such as the Kuroshio in the Pacific and the Gulf Stream in the Atlantic. Where these flows are most vigorous, they can extend to great depth and even to the ocean bottom.

How do the winds drive this flow? The winds push on the very top of the ocean, and move the water through frictional stress. This frictional layer is referred to as the "Ekman layer" and is a total of about 20 to 100 meters deep. The resulting movement of the water is to the right of the wind in the northern hemisphere and to the left in the southern hemisphere. This very thin water layer then pushes on a thin water layer below it through friction, and so on. Each of the thin water layers pushes the one below it slightly more to the right (northern hemisphere). The frictional stress becomes smaller and smaller with depth as the energy is put into moving the water. In fact the frictional stress dies out at about 50 meters below the sea surface. Thus the winds frictionally drive only the very top of the ocean. The overall effect of the wind on this 50-meter layer is to drive a net flow of water exactly to the right of the wind (northern hemisphere). This is called "Ekman transport". It adds on to the geostrophic surface flow, which, as said above, is driven by a pressure difference. Using surface drifters which report their positions via satellite, Ralph and Niiler have mapped the average flow at 15 meters depth in the Pacific. When they subtract the fairly well-known large-scale geostrophic flow from their average flow, the resulting flow is indeed to the right of the wind in the northern hemisphere and to the left in the southern hemisphere, which substantiates the idea of Ekman transport on a large spatial scale.

Winds are highly variable in general - weather patterns come and go in a matter of days. However, winds averaged over a season or a year or many years drive the large-scale, slowly-changing ocean circulation. Because the average winds vary in strength and also in direction over a large scale, the surface layer Ekman flow varies in strength and direction. Where the surface flow converges (flows together), there must be downwelling, and where it diverges (flows apart), there must be upwelling.

In regions of downwelling, the ocean lying beneath the surface layer, and down to about 2000 meters depth, responds with slow equatorward flow. The reason for this equatorward flow is more or less angular momentum conservation - as a vertical column of water that rotates due to the earth's rotation is squashed by downwelling, it must spin more slowly. To spin more slowly it moves towards the equator where the amount of earth's rotation which projects onto the vertical column is lower. Such slow equatorward flow is found in the subtropics (20° to 40° from the equator) where there are westerlies at higher latitude and easterly trades at lower latitude. This slow equatorward flow is fed from the western boundary, by eastward flow at latitudes of 30° to 50°. The equatorward flow returns to the western boundary at latitudes of about 15° to 30° in the northern hemisphere and from 30°S to the equator in the southern hemisphere. The western boundary current which feeds this circulation in the northern hemisphere is the Kuroshio. In the southern hemisphere it is the East Australia Current.

In regions of upwelling, the underlying ocean flow is poleward, away from the equator. This occurs at high latitudes (greater than 50°) and also in the narrow band under the Intertropical Convergence Zone at about 5° to 10°N. The result is a counterclockwise circulation. In the northern North Pacific, the western boundary current which feeds this gyre is the Oyashio. In the tropics north of the equator, the currents are nearly due east-west but they do have a slight counterclockwise gyre configuration and a western boundary current. The currents in this tropical gyre are the North Equatorial Countercurrent, which flows eastward on the southern side of this cell, and the southern part of the North Equatorial Current, which flows westward on the northern side of this cell. Its western boundary current is the Mindanao Current.

Wind-driven Circulation at the Equator

Directly on the equator, the effect of rotation on the circulation vanishes, and so these concepts of geostrophic and Ekman flow do not apply. At the equator, the easterly trade winds push the surface water directly from east to west. This water piles up gently in the western Pacific (0.5 meters higher there than in the eastern Pacific). Because it is higher in the west than in the east, there is a pressure difference which causes the flow just beneath the surface layer to be eastward. This strong eastward flow is the Equatorial Undercurrent.

The alternating eastward and westward jets found below the Equatorial Undercurrent on the equator die out about 2° from the equator. Their cause has not been clearly identified. However the theory of very slow waves on the equator, which move water from side to side much more than the up and down of surface waves, shows us that equatorial waves have much more complicated (reversing) vertical structure than waves off the equator. It is expected that this complex structure for the very slow waves with very long east-west wavelengths translates to complex structure in the mean currents.

Also occurring very close to the equator is northward Ekman transport north of the equator and southward Ekman transport south of the equator, due to the easterly trade winds (blowing from east to west). This causes upwelling right at the equator.

Along the equator just below the surface, waters in the east are colder than in the west. This is partially a result of the rising of the Equatorial Undercurrent from west to east in response to upwelling. Upwelling in the eastern Pacific thus accesses much cooler water than in the western Pacific, and as a result the surface waters in the east are colder than in the west. Steady trade winds, which cause equatorial upwelling, are more prevalent in the east than in the west. There is seasonality in the winds, and equatorial upwelling is weaker in the northern winter and spring, giving rise to mini-El Nino conditions each year in the eastern equatorial Pacific.

Response to Changing Winds in the Tropics

When the trade winds weaken or even reverse, the flow of water westward at the equator weakens or reverses and upwelling weakens or stops. Surface waters in the eastern Pacific warm significantly since upwelling is no longer bringing the cool waters to the surface. The deep warm pool in the western Pacific thins as its water sloshes eastward along the equator in the absence of the trade winds which maintain it.

Heating/Cooling and Evaporation/Precipitation

Ocean water density is a function of temperature and salinity, and so can be changed through heating/cooling and evaporation/ precipitation. The resulting density changes can drive circulation, but density-driven flow is much weaker than that driven by the winds. However, density changes, caused mainly by fluxes at the sea surface, are the only means of forcing circulation where the indirect effect of wind forcing vanishes, as in the ocean deeper than about 2000 meters. In the upper ocean, even though density fluxes do not greatly change the flow, they do have a major effect on ocean properties and on the overlying atmosphere, which is heated from below by the ocean. The total surface heat flux into the ocean averaged over all years of data shows the greatest heating along the equator and in the western warm pool region around Indonesia. The units of heating are

Watts/m², or energy per unit area. The uncertainty in heating is about 20 Watts/m² and so values lower than this are not significantly different from zero. In the subtropics where the western boundary currents bring warm water to mid-latitudes, there is strong cooling. In order to maintain a fairly steady distribution of temperature, the ocean must transport heat from the areas where it gains heat to the areas where it loses it. The large arrows in figure show the direction of heat transport in each ocean basin across the latitudes where the arrows are placed.

In the western warm pool region and all along the ITCZ there is major convection in the atmosphere, creating towering clouds. Precipitation in these regions creates pools of freshened surface waters. At mid- latitudes, excess evaporation under the atmospheric high pressure cells creates high salinity surface water. These waters can be traced by their salinity as they move to below the sea surface and are carried far by the ocean currents.

Paleoceanography

Paleoceanography is the study of the history of the oceans. It encompasses aspects of oceanography, climatology, biology, chemistry and geology. The main sources of information are biogenic and inorganic marine sediments, as well as corals. Biogenic sediment includes planktonic and benthic fossils whereas inorganic sediment includes ice-rafted debris and dust. On land, paleo-shorelines and erosional features as well as outcroppings of paleomarine sediments are the principal sources of proxy data. Glaciological records can also give indirect information about paleoceans. The ocean's high heat capacity and its ability to transport energy and to sequester and release greenhouse gases give it an important role in helping to determine the state of the planet's climate. Thus, paleoceanographic research is also intimately linked to paleoclimatology.

Methods

The reconstruction of paleocean characteristics and dynamics requires climatic detective work. It involves the dating and interpretation of paleoclimatic records as well as the definition of physical and dynamical constraints which specify possible circulation patterns and characteristics.

Reconstructions (Proxy Data)

Direct measurements of the quantities of interest to oceanographers extend only into the relatively recent past and in most cases do not go further back than the mid-nineteenth century. To study the ocean during periods for which there are no direct measurements one must rely on indirect evidence. Historical documents can be used as sources of data. Ship logbooks and sailing times across frequently traveled routes have provided estimates of the directions and strengths of past prevailing winds. The frequency and intensity of El Niño events since the 1500s have been reconstructed based partially on historical accounts of large floods and crop losses. This type of analysis furnishes qualitative descriptions of the past.

Quantitative reconstructions are possible by proxy, where a quantity which is preserved in a natural archive and can be measured, stands as a surrogate of the parameter of interest. A basic requirement is that the relationship between the proxy parameter and the quantity of interest has to be known. When this is the case, the history of the proxy variable can be converted into the history of the variable of interest by the use of mathematical expressions of the type:

$$Int_t = f\left(\Pr x_t\right)$$

Which state that the parameter of interest, Int, is a function of the proxy quantity, Prx. The t index refers to time. Equations of this kind are commonly called transfer functions. The confidence with which Int can be estimated will depend on a series of factors, starting with the quality of the proxy measurements. Also relevant, and a common source of uncertainty, is how well f represents the relationship between Prx and Int.

In most cases, transfer functions are obtained empirically by comparing directly measured values of the quantity of interest to a pertinent set of proxy data. A potential source of error is that the function obtained by this procedure might not be general, but in fact could represent a relationship between Prx and Int that is peculiar to the data sets used to generate f. This problem can be minimized by expanding the spatial and temporal coverage of the data used to establish the transfer function. Still, even assuming that f is a perfect representation of how the proxy and the quantity of interest are connected to each other in the present, there is no guarantee that the relationship between them was the same in the past.

Another source of error can be easily understood by rewriting Equation $Int_t = f\left(\Pr x_t\right)$ so that it expresses the proxy quantity as a function of the variable of interest. It is reasonable to assume (and in many cases it has been demonstrated) that Int is not the only factor controlling Prx, so that in fact we end up with:

$$\Pr x_t = f^{-1}\left(Int_t, E_{1t}, E_{2t}, ...E_{nt}\right)$$

Where, E_1, E_2,......E_n represent environmental parameters that also influence the proxy variable but are independent of the quantity of interest. An immediate conclusion is that reconstructions of Int based on Prx will be "contaminated" by other factors so that part of the variability observed in the proxy quantity is not related to changes in the parameter of interest. Comprehensive analysis of the relationships between proxies and a series of observed parameters can offer some insight into how to remove part of the undesired influence of other factors from the reconstruction.

Given the complexity involved in developing skillful transfer functions as well as in identifying and correcting for potential sources of error, a common strategy is to reconstruct the same parameter of interest using different proxies. Such analyses are known as multi-proxy reconstructions.

Types of Proxies

The systematic use of proxies in quantitative reconstructions of past oceanic environments originated in the second half of the twentieth century. Since then, a large number of proxy techniques have been established and more are constantly being developed. Proxies can be grouped in six broad categories, based on the type of direct measurement. These are listed below, together with brief descriptions of the main variables of interest associated with each proxy:

- Microfossil assemblages: The relative abundance of planktonic and benthic species of foraminifera, coccoliths, radiolaria, diatoms and other organisms can be used to estimate past ocean temperature, productivity and sea ice distribution. This proxy type was used for the CLIMAP project, which produced the first global distribution of sea surface temperature for the Last Glacial Maximum.

- Stable isotopes are based on the ratio between different isotopes of an element: The ratios are usually standardized by a reference value and named after the heavier isotope. The ratio between ^{18}O and ^{16}O, for example, is represented by $\delta^{18}O$. Isotope readings are retrieved mainly from foraminifer's skeletons (tests), organic matter or other sources (e.g., water molecules in continental ice sheets). The amount of ^{18}O incorporated by organisms like foraminifera and corals increases as temperature decreases. Continental ice is relatively depleted in ^{18}O compared to sea water. This makes $\delta^{18}O$ a proxy for both temperature and the extent of continental ice sheets. $\delta^{11}B$ is used as a proxy for pH. Productivity, nutrient concentration and past circulation can be reconstructed from $\delta^{15}N$ and $\delta^{13}C$ (^{12}C is taken up with slight preference to ^{13}C during photosynthesis). Together with microfossil assemblages, $\delta^{18}O$ and $\delta^{13}C$ are the paleoceanographic proxies with the most widespread use.

- Radiogenic isotopes: The different solubilities of uranium and two products of its naturally occurring decay, thorium (Th) and protactinium (Pa) can be used to estimate the rate of deep water flow and the flux of particles from the water column to the sediments. This flux can also be used as a productivity estimate. ^{14}C preserved in organic matter is used to estimate the age difference between near surface and deep waters and hence, ventilation rates.

- Biogenic compounds: The concentrations of some compounds, mainly organic carbon, calcium carbonate and opal, are used as estimates of past productivity. Calcium carbonate is also an indicator of the calcite compensation depth. Alkenones are long chained organic molecules resistant to degradation. The alkenones produced by some coccolithophors can have two or three double bonds in their structure. The ratio between molecules with two and three double bonds reflect the temperature at the time of synthesis.

- Elements: The concentrations and ratios of certain elements in the sediment, organic remains, tests and corals are also used as proxies. The ratios of strontium to calcium (Sr/Ca) and magnesium to calcium (Mg/Ca) in biologically precipitated marine carbonates are temperature dependent. The cadmium to calcium (Cd/Ca) ratio is used for nutrient reconstructions. Barium concentration and the Ba/Ca ratio are proxies for productivity and alkalinity, respectively.

- Sedimentology: Grain size distribution can provide qualitative information about bottom current speeds and act as an indicator of ice rafted debris. Information about past tides can be inferred from layered sediments called rhythmites. The mineralogy of the sediments can be used to establish source areas and direction of transport for both water and wind borne sediments.

Conspicuously absent from the quantities of interest listed above is salinity, a fundamental parameter which influences many aspects of the ocean environment. There is, at the moment, no independent proxy for this quantity. As salinity also influences $\delta^{18}O$, an indirect measurement can be obtained by using independent estimates of temperature (alkenones, for example) to remove the temperature signal from existing $\delta^{18}O$ series. Attempts have also been made to infer salinity from microfossil assemblages. Unfortunately, both approaches generate errors of ~1 psu, very large compared to the range of salinity variability in the oceans.

Reconstructions Models

The theoretical approach to paleoceanography uses quantitative ocean and climate models to reconstruct paleoconditions of the ocean and interpret observations. A number of different ocean and coupled climate models have been used in the past and we will give a very short overview of the hierarchy and use of these models in the broad research area of paleoceanography.

It is, of course, impossible to build an exact model of the climate system; for the ocean alone, the position and momentum of approximately 5×10^{46} molecules would have to be calculated at each instant of time. As an alternative, the ocean, atmosphere, sea ice, or land surface is split into discrete macroscopic elements with measurable characteristics such as temperature, density, velocity, etc. The state of and exchange between these discrete elements follow physical laws and can therefore be determined with numerical models. Small scale processes within each element can also influence the large scale pattern and therefore have to be parametrized. Existing ocean (and climate) models differ in regards to:

- Their temporal and spatial resolution (resolution is defined as the spatial scale which defines the boundary between processes that are resolved by the model and those which are parametrized).

- The nature of processes which are resolved (for example, some models include biogeochemical cycles whereas other models resolve physical processes only).

- The number of subsystems taken into account (for example, an ocean model needs boundary conditions at the surface which can either be provided by data, or by an atmosphere model which is physically interacting with the ocean model).

As the different subsystems (ocean, atmosphere, sea ice, continental ice sheets, vegetation, etc.) of the climate system interact with each other in a complex manner and on a very broad range of timescales, climate modelling reduces to the process of identifying isolable subsystems and processes that are relevant to the problem at hand. While identifying these subsystems and processes, the researcher has also to keep in mind that these processes have to be suitable to be simulated by limited mathematical models and will provide results in a reasonable computational time.

The simplest class of climate models includes one-dimensional Energy Balance Models which were first developed by Budyko and Sellers. It is interesting to note that these simple models yield two stable solutions under present day boundary conditions: the present day climate and a completely frozen Earth (also called "Snowball Earth").

Today's physical ocean models can be classified based on the following categories:

- Geography (regional models, global models).

- Physics (hydrodynamic, thermodynamic or hydrothermodynamic models).

- Surface approximation (free surface, rigid lid).

- Vertical discretization (fixed level, isopycnal, sigmacoordinate, semi-spectral).

- Density variation (barotropic, baroclinic).

Because the boundary data for paleoclimatic simulations tend to contain large uncertainties, global models are better suited for this research area than regional models. The most comprehensive results are given by global ocean general circulation models (OGCMs). These models can be driven with reconstructed data specifying the boundary conditions. For example, numerous modelling studies restored the ocean surface characteristics to the CLIMAP data set (CLIMAP – Project Members, 1976) for simulations of the Last Glacial Maximum.

A better approach than using ocean-only models is to use coupled ocean-atmosphere models; by computing the surface boundary conditions, one can bypass the data problem. However, coupled ocean-atmosphere GCMs often need flux adjustments. Flux adjustments balance surface fluxes at the ocean-atmosphere interface to avoid a numerical drift of the coupled system. As flux adjustments have been "tuned" to the present day climate, the use of these adjustments to simulate past climates is not very reliable. However, some recent studies use coupled atmosphere-ocean GCMs which do not need artificial flux adjustments.

Some studies use simple atmosphere models which still provide reasonable boundary conditions for the ocean model. Other climate subsystems may also have an important influence on the state of the ocean (continental ice sheets, sea ice and land surface processes for example). The class of Earth System Models has been developed recently and comprise all models which take into account more than two subsystems of the climate system. Today, Earth System Models are widely used for paleoceanographic and paleoclimate simulations. There is also growing evidence that geochemical interactions between the subsystems are more important than initially thought (carbon cycle, nitrogen cycle, methane, etc.) and some initial attemps in integrating these cycles in Earth System Climate models for paleoclimatic simulations have been made. The modelling community is continuously integrating new processes and subsystems in their models to obtain a better representation of the climate system dynamics.

Combining Models and Proxy Data

The interpretation of paleoproxy data is an ongoing challenge for paleoclimate scientists. As a striking example, one could compare the studies of Clark et al. and Bond et al. Both papers are highly regarded, yet draw opposite conclusions from the atmospheric $\Delta^{14}C$ record. Whereas Bond et al. relate the variability of atmospheric $\Delta^{14}C$ to changes in solar radiation, Clark interpret the same type of record as a signature of variability in the thermohaline circulation and ocean heat transport. On the other hand, model simulations of past climates depend strongly on boundary conditions, assumptions, and the model used for the study. The simulated climate for a certain time span can be radically different depending on the model and boundary conditions used. Interpretation of measured paleoclimate data is thus urgently needed through collaboration between modellers and observers.

One possible approach to combine proxy data with climate models is that of "data assimilation." Data assimilation involves the construction of a field that accommodates best the information obtained from paleoproxies with the physical (and dynamical) constraints of the climate system using coupled climate models.

Another possible approach is to incorporate paleoproxy data (e.g., $\delta^{18}O$, deuterium excess, $\Delta^{14}C$, $\delta^{13}C$, $\delta^{10}Be$, etc.) as prognostic active tracers in climate models. Perturbations (such as meltwater events or changing solar activity) and other climate states (e.g., the Last Glacial Maximum) can then be simulated and the behaviour of these simulated proxies can be compared to observed proxy data obtained from ice cores, marine sediments and other records. This has been done with uncoupled ocean (or atmosphere) general circulation models, vegetation models and continental ice sheet models. However, the importance of interactions between atmosphere, oceans and other systems such as the biosphere and the cryosphere point to the necessity of using coupled models. To date, there have been only a few studies simulating paleoproxy data with either coupled ocean-atmosphere GCMs or Earth System Models.

A third way to bridge the gap between the modelling and proxy data approach is to find locations of proxy data records of special interest with the use of climate models. A simulation including prognostic paleoproxy tracers can determine the geographical region of greatest impact on a given paleoproxy data during a given climate event.

In conclusion, large amounts of paleoproxy data have been retrieved from various types of archives, but attempts to use numerical models for verification and interpretation of this data are sparse. The science of using three-dimensional climate models to interpret paleo records is still in its infancy.

Processes

Paleotides

The history of tides over geological time is associated with the evolution of the Earth-Moon system and the shape of ocean basins. Tidal currents generate friction at the bottom of the ocean resulting in the transfer of energy and angular momentum associated with the Earth's rotation to the Moon's orbital motion. This process has been gradually slowing the Earth's spin and increasing the radius of the Moon's orbit. According to some estimates, at ~620 million years ago (Ma), days were approximately 22 h long and the Earth-Moon distance was about 96% of its present day value. The main proxy for tidal (and Earth-Moon system) changes over periods of millions of years is based on the analysis of tidal rhythmites, laminated sediments whose deposition is associated with tidal currents.

Tidal dissipation depends strongly on the shape of ocean basins. As tectonic processes cause the basins to change, the effects of tidal friction should also change. There are many indications that this is the case. For example, the present rates of dissipation appear to be higher than the average rates over the planet's history.

Over the Quaternary, the shape of ocean basins was altered due to changes in sea level and the presence or absence of ice shelves. Numerical models show that in the Labrador Sea, tidal amplitudes during the last glacial were about twice as large compared to present day conditions. This has led to the suggestion that these higher tides could have destabilized floating ice shelves and caused Heinrich events. Another connection between tides and climate relates to a millennial tidal cycle. It has been proposed that very high tides, occurring every 1,800 years, can cause increased mixing and cooling of the sea surface. This cooling would be related to abrupt climate change observed with similar periodicity. Some researchers contest this hypothesis, arguing that tidal forcing at these frequencies is very weak.

Radiation

Incoming short wave radiation from the Sun is the ultimate source of energy for ocean dynamics, the hydrological cycle and life in the oceans. Geographic and seasonal variations in the intensity of insolation result in temperature and pressure gradients which have an important influence on ocean dynamics and the climate system. Although the ocean surface circulation is mostly wind driven, the winds themselves are the result of the uneven distribution of energy on Earth. Density driven currents on the other hand depend strongly on temperature (and thus indirectly on energy distribution) and salinity gradients (which result from precipitation/evaporation patterns and are thus closely linked to the hydrological cycle).

Incoming solar radiation changes over a range of very different timescales. Firstly, the solar luminosity has gradually increased throughout the Earth's lifetime. It is estimated that, during the early days of our planet, the solar luminosity was 25–30% weaker than today's value. According to model results, such a reduction in incoming shortwave radiation should result in a completely frozen planet ("Snowball Earth"). However, even if evidence exists for snowball Earth conditions in early Earth's climate history, there is also counter evidence that during long periods of time the paleoceans were ice-free. Thus, it is inferred that, at these times, greenhouse gas concentrations must have been higher than present day levels in order to prevent the system from slipping into an icehouse state.

On shorter timescales (order of tens to hundred thousands of years), the incoming solar radiation is modulated by changes in the Earth's orbital parameters which describe the character of the Earth's orbit around the Sun. These three parameters change continuously, causing a variation in the total amount of solar radiation received on Earth as well as the seasonal and latitudinal distribution of insolation. The first parameter, called eccentricity, describes the degree of ellipticity in the Earth's orbit around the Sun and hence the shape of its orbit. The characteristic periods of changes in eccentricity are 95,000, 131,000, 413,000, and 2,100,000 years. Eccentricity is the only parameter which modulates the total global amount of solar energy received at the top of the atmosphere. The change in Earth's axial tilt through time is described by obliquity and has a distinct period of 41,000 years. Finally, the precession of the equinoxes combines the axial precession (wobbling of the axis) and the precession of the ellipse (rotation of the elliptical shape of Earth's orbit) and consists of a strong cycle with a 23,000 year period and a weaker one with a 19,000 year period. Precession and obliquity do not alter the total amount of solar radiation received; however, they change the distribution of incoming radiation by latitude and by season. All the frequencies of these parameters can be found in climatological records of the paleocean. Thus, orbital parameters play an important role in driving the climate system and ocean dynamics.

Finally, short-term variability in solar luminosity (which is correlated with changes in the number of sunspots visible on the Sun's surface) acts on timescales of decades to millenia. Over the last hundred years, the global mean temperature has followed a trend similar to the sunspot record. Some climate scientists have hypothesized those periods of global cooling.

Ocean Basin Changes

The shape of the ocean basins sets the boundary conditions for the ocean circulation. Different continent configurations result in different flow patterns. Understanding how the boundaries influenced past circulation patterns might offer insight into the dynamics and other important processes of the present day oceans. For example, the existence of an unobstructed low latitude passage in the Tethys Ocean (160–14 Ma) has been associated with increased poleward heat transport. Of course, both reconstructions and simulations of ocean currents this far into the past are subject to many uncertainties. For example, even the existence of a prominent, large scale feature such as the Tethys circumequatorial current is still not unequivocally accepted.

A number of proxy and modelling studies show that the opening and closing of passages between two basins can have a large impact on ocean circulation and climate. The closing of the isthmus of Panama which is the gap between the North and South American continents at 5–3 Ma is thought to have intensified the Gulf Stream and the associated northward transport of heat and salt into the North Atlantic. The closure of the isthmus of Panama has also been associated with the onset of Northern Hemisphere glaciation as well as with changes in water mass properties and the overturning circulation of the North Pacific. The opening of the Drake passage between Antarctica and South America (28–33 Ma) and subsequent establishment of the Antarctic Circumpolar Current is associated with the glaciation of the Antarctic continent. It has been proposed that the relative stability of the Atlantic Meridional Overturning Circulation (MOC) during the Holocene is related to the open connection between the Arctic and Pacific Oceans provided by the Bering Strait. According to this scenario, low salinity anomalies in the North Atlantic make the MOC unstable when the strait is blocked (as during the last glacial). On the other hand, the flux of low salinity water from the Pacific into the Arctic through a wide open Bering Strait during conditions with higher sea levels (last interglacial, 115–130 ka BP) has been linked to the more unstable MOC.

Oscillations in sea level result in changes of the surface's land to ocean ratio which in turn influences the planetary albedo. According to model results, alteration of the planet's albedo caused by a sea level drop of approximately 400 m during the late Ordovician (455–445 Ma) was one of the factors that could have driven the climate into a cooler state. By submerging vegetated areas or making new land available to plants, changes in sea level can also impact carbon storage. The amount of carbon present on shelves inundated by the rise in sea level since the Last Glacial Maximum appears to be equivalent to the increase in the atmospheric stock of carbon during the same period.

Ice

A gradual cooling over the last 55 million years led to the presence of extensive continental ice sheets in both the Southern and Northern Hemisphere. Intensification of Northern Hemisphere glaciation in Eurasia and North America occurred between 2.7 and 2.5 Ma. With increasing terrestrial ice sheets, sediment records show an intensification of climate oscillations between extreme cold glacial maxima and interglacial warm periods on timescales related to the orbital parameters. The boundary conditions for paleoceans varied dramatically between glacial and interglacial periods. The formation and melting of extensive ice sheets in North America and Eurasia, important variations in atmospheric CO_2 concentrations as well as shifts in the extent of sea ice changed the radiation balance, salinity distribution, dynamical forcing and productivity of the oceans.

Continental Ice Sheets

Continental ice sheet growth and decay during glacial cycles may have influenced the ocean circulation in several ways:

- Sea level changes due to the storage of freshwater on continents and bedrock depression may have led to different circulation patterns in the ocean.

- Extraction of freshwater from the ocean to build up continental ice sheets led to changes in the global average and distribution of ocean salinity. These changes in salinity (and therefore density) may in turn have influenced the circulation and heat transport.

- Changes in land surface albedo altered the local radiation balance over continental ice sheets, which in turn must have changed sea surface temperatures and circulation.

- Elevation changes in regions of continental ice sheets could also have caused changes in atmospheric circulation by affecting the atmosphere's stationary wave pattern and therefore the dynamic forcing of the ocean circulation.

- Changes in global atmospheric temperatures resulting from the presence or absence of ice sheets led to changes in ocean temperatures which may in turn have affected the circulation and heat transport.

Climate variability on millenial time scales has been more important during glacial periods than during interglacials. Dansgaard-Oeschger temperature oscillations, which seem to be part of slow-cooling cycles occurring every 10–15 kyr, have been related in several studies to changes in the strength of the meridional overturning. Local sea surface salinity perturbations at high latitudes due to meltwater and iceberg discharges (Heinrich Events) may have weakened the MOC leading to abrupt climate change.

Sea Ice

Sea ice regulates exchanges of heat and freshwater between ocean and atmosphere and can change sea surface salinity through melting or freezing (brine rejection). It insulates the relatively warm ocean water from the cold polar atmosphere, changes the surface albedo (and therefore the local radiation balance) dramatically and influences evaporation and therefore local cloud cover and precipitation. By changing the surface characteristics of the oceans, sea ice plays an important role in deep water formation and meridional heat transport in the ocean. The positive ice-albedo climate feedback might have led to a "run away" icehouse feedback in the Precambrian.

References

- Biological-oceanography, studies: phdportal.com, Retrieved 26 July, 2019

- Biological-oceanography: mit.whoi.edu, Retrieved 21 May, 2019

- Abraham; et al. (2013). "a review of global ocean temperature observations: implications for ocean heat content estimates and climate change". Reviews of geophysics. 51 (3): 450–483. Citeseerx 10.1.1.594.3698. Doi:10.1002/rog.20022

- Oceanography-biological, oc-po: waterencyclopedia.com, Retrieved 8 January, 2019

- Jenkins, adrian; et al. (2016). "decadal ocean forcing and antarctic ice sheet response: lessons from the amundsen sea | oceanography". Tos.org. Retrieved 2019-02-05

- Ocean-biogeochemistry, news-wires-white-papers-and-books, science: encyclopedia.com, Retrieved 25 February, 2019

- Ocean-acidification, science: britannica.com, Retrieved 13 May, 2019

- Geological-oceanography, oceanography: scitechnol.com, Retrieved 16 January, 2019

Chapter 3

Ocean Ecosystems

The ocean ecosystem is a vast area of study. It can broadly be classified into open ocean ecosystem, deep ocean ecosystem, coral reef ecosystem, shoreline ecosystems, etc. These diverse types of ocean ecosystems have been carefully analyzed in this chapter.

Ocean ecosystems or marine ecosystems are made up of a community of living and non-living things found in a localized area in any ocean. In an ecosystem, the plant life and animal life support one another and are each dependent on the other for the success of the ecosystem.

According to National Geographic, there are many ocean ecosystems, including the abyssal plain, Polar Regions, coral reefs, the deep ocean, mangroves, kelp forests, salt marshes and sandy shores, among others. The types of plants and animals found in each ecosystem are distinct and are suited for life in their respective ecosystems.

Broadly speaking, the marine ecosystem refers to the oceans and seas and other salt water environments as a whole; however, it can be divided into smaller, distinct ecosystems upon closer inspection. There are various types of marine ecosystems, including salt marshes, estuaries, the ocean floor, the broad ocean, the inter-tidal zones, coral reefs, lagoons, and mangroves.

In accordance with, but not necessarily because of, their large size and wide range, marine ecosystems are also easily the most diverse of all the ecosystems on the planet. Coral reefs alone are home to over 25% of all marine life, despite occupying less than 1% of the ocean floor.

Like all ecosystems, marine ecosystems are finely balanced and highly complex. There are many different parts that make up an ecosystem, and each part plays a role in maintaining balance within the system. Organisms depend on, and are highly influenced by, the physiochemical environmental conditions in their ecosystem.

Marine Ecosystem Food Chain

A food chain refers to a series of organisms that are interrelated in their feeding habits. It is hierarchical in manner, with smaller organisms being fed upon by larger organisms, which in turn feed even larger organisms, and so on. All food chains begin with a producer, which is consumed by a primary consumer, which is consumed by a secondary and then tertiary consumer, and ultimately maps the flow of energy throughout trophic levels. A marine ecosystem food chain is a food chain that is specifically found within marine ecosystems.

Many marine food chains begin with phytoplankton. Phytoplankton are microscopic marine algae no bigger than 20mm. Although seemingly insignificant, phytoplankton provide the foundation, or the first level, of the sea's food chain in a balanced system. In areas where there is enough light to support photosynthesis, such as the upper parts of the ocean's surface, these microalgae provide two important services. First, they serve to produce approximately 50% of the world's oxygen. Second, they support the populations of primary consumers that feed on them and, indirectly, the populations of higher level consumers that feed on them.

This image depicts a food web with multiple possible food chains with the arrows pointing in the direction of energy flow. In the sea, all arrows lead back to the phytoplankton in the lower right-hand corner. The phytoplankton produce their own energy via photosynthesis. From there, we can see the flow of energy going toward the primary consumers that eat them, such as the school of small fish. The energy is passed on again when those fish are consumed by the larger fish, and passed on again when that fish is consumed by the shark. In this food web, the shark is the top tertiary consumer. The energy in the shark is cycled back into the ecosystem only when the shark dies and its body is consumed by detritivores.

This is just one example of a food chain found in the marine ecosystem. Another food chain might begin with seaweed being eaten by sea urchins. Another still might start with sea grass being eaten by sea turtles.

One thing to remember is that only 10% of energy is passed on from one trophic level to the next, meaning that higher level predators need to consume many lower level prey to sustain themselves. Because of this, there will always need to be a higher number of lower level trophic organisms than higher level predators in an ecosystem.

Basic Marine Ecosystem Facts

Marine Ecosystem Animals

Marine ecosystems support a great diversity of life with a variety of different habitats. They can be categorized into groups based on where they live (benthic, oceanic, neritic, intertidal), as well as by shared characteristics (vertebrates, invertebrates, plankton). Specific examples of marine animals include sea urchins, clams, jellyfish, corals, anemones, segmented and non-segmented worms, fish, pelicans, dolphins, phytoplankton, and zooplankton.

Marine Ecosystem Plants

You can find many types of plants in the ocean, including seaweeds, algae (red, green, brown), sea grasses (the only flowering plants in the marine ecosystem), and mangroves.

Marine Ecosystem Climates

Marine ecosystems are found on many different parts of the Earth, so it shouldn't be surprising to learn that marine climates can vary from tropical to polar. Other climates found in marine ecosystems include monsoon, subtropical, temperate, and subpolar.

Marine Biota

Marine biota can be classified broadly into those organisms living in either the pelagic environment (plankton and nekton) or the benthic environment (benthos). Some organisms, however, are benthic in one stage of life and pelagic in another. Producers that synthesize organic molecules exist in both environments. Single-celled or multicelled plankton with photosynthetic pigments are the producers of the photic zone in the pelagic environment. Typical benthic producers are microalgae (e.g., diatoms), macroalgae (e.g., the kelp Macrocystis pyrifera), or sea grass (e.g., Zostera).

Plankton

Plankton are the numerous, primarily microscopic inhabitants of the pelagic environment. They are critical components of food chains in all marine environments because they provide nutrition for the nekton (e.g., crustaceans, fish, and squid) and benthos (e.g., sea squirts and sponges). They also exert a global effect on the biosphere because the balance of components of the Earth's atmosphere depends to a great extent on the photosynthetic activities of some plankton.

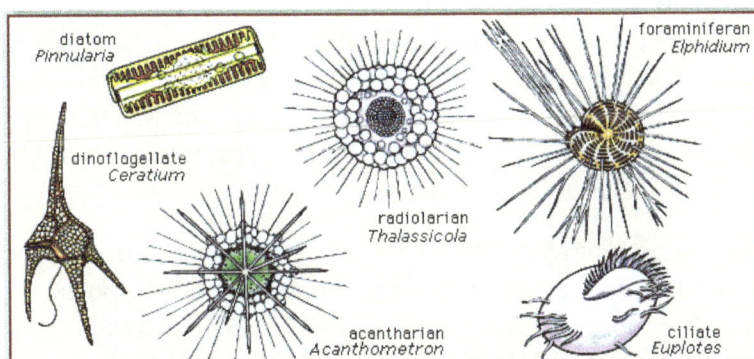

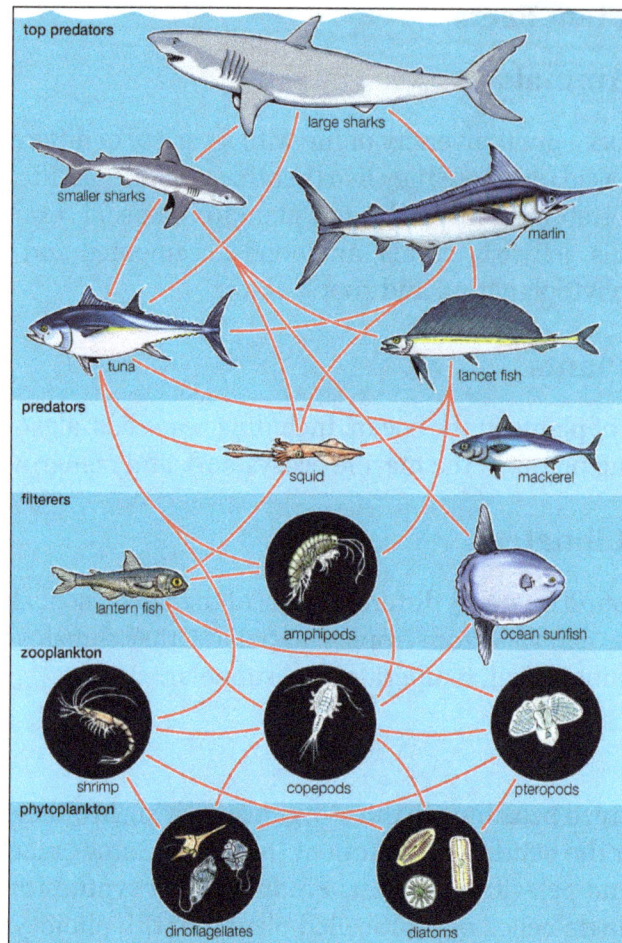

The meaning of term plankton is wandering or drifting, an apt description of the way most plankton spend their existence, floating with the ocean's currents. Not all plankton, however, are unable to control their movements, and many forms depend on self-directed motions for their survival.

Plankton range in size from tiny microbes (1 micrometre [0.000039 inch] or less) to jellyfish whose gelatinous bell can reach up to 2 metres in width and whose tentacles can extend over 15 metres. However, most planktonic organisms, called plankters, are less than 1 millimetre (0.039 inch) long. These microbes thrive on nutrients in seawater and are often photosynthetic. The plankton include a wide variety of organisms such as algae, bacteria, protozoans, the larvae of some animals, and crustaceans. A large proportion of the plankton are protists—i.e., eukaryotic, predominantly single-celled organisms. Plankton can be broadly divided into phytoplankton, which are plants or plantlike protists; zooplankton, which are animals or animal-like protists; and microbes such as bacteria. Phytoplankton carries out photosynthesis and are the producers of the marine community; zooplankton are the heterotrophic consumers.

Diatoms and dinoflagellates (approximate range between 15 and 1,000 micrometres in length) are two highly diverse groups of photosynthetic protists that are important components of the plankton. Diatoms are the most abundant phytoplankton. While many dinoflagellates carry out photosynthesis, some also consume bacteria or algae. Other important groups of protists include flagellates, foraminiferans, radiolarians, acantharians, and ciliates. Many of these protists are important consumers and a food source for zooplankton.

Zooplankton, which are greater than 0.05 millimetre in size, are divided into two general categories: meroplankton, which spend only a part of their life cycle—usually the larval or juvenile stage—as plankton, and holoplankton, which exist as plankton all their lives. Many larval meroplankton in coastal, oceanic, and even freshwater environments (including sea urchins, intertidal snails, and crabs, lobsters, and fish) bear little or no resemblance to their adult forms. These larvae may exhibit features unique to the larval stage, such as the spectacular spiny armour on the larvae of certain crustaceans (e.g., Squilla), probably used to ward off predators.

Important holoplanktonic animals include such lobsterlike crustaceans as the copepods, cladocerans, and euphausids (krill), which are important components of the marine environment because they serve as food sources for fish and marine mammals. Gelatinous forms such as larvaceans, salps, and siphonophores graze on phytoplankton or other zooplankton. Some omnivorous zooplankton such as euphausids and some copepods consume both phytoplankton and zooplankton; their feeding behaviour changes according to the availability and type of prey. The grazing and predatory activity of some zooplankton can be so intense that measurable reductions in phytoplankton or zooplankton abundance (or biomass) occur. For example, when jellyfish occur in high concentration in enclosed seas, they may consume such large numbers of fish larvae as to greatly reduce fish populations.

The jellylike plankton are numerous and predatory. They secure their prey with stinging cells (nematocysts) or sticky cells (colloblasts of comb jellies). Large numbers of the Portuguese man-of-war (Physalia), with its conspicuous gas bladder, the by-the-wind-sailor (Velella velella), and the small blue disk-shaped Porpita porpita are propelled along the surface by the wind, and after strong onshore winds they may be found strewn on the beach. Beneath the surface, comb jellies often abound, as do siphonophores, salps, and scyphomedusae.

The pelagic environment was once thought to present few distinct habitats, in contrast to the array of niches within the benthic environment. Because of its apparent uniformity, the pelagic realm was understood to be distinguished simply by plankton of different sizes. Small-scale variations in the pelagic environment, however, have been discovered that affect biotic distributions. Living and dead matter form organic aggregates called marine snow to which members of the plankton community may adhere, producing patchiness in biotic distributions. Marine snow includes structures such as aggregates of cells and mucus as well as drifting macroalgae and other flotsam that range in size from 0.5 millimetre to 1 centimetre (although these aggregates can be as small as 0.05 millimetre and as large as 100 centimetres). Many types of microbes, phytoplankton, and zooplankton stick to marine snow, and some grazing copepods and predators will feed from the surface of these structures. Marine snow is extremely abundant at times, particularly after plankton blooms. Significant quantities of organic material from upper layers of the ocean may sink to the ocean floor as marine snow, providing an important source of food for bottom dwellers. Other structures that plankton respond to in the marine environment include aggregates of phytoplankton cells that form large rafts in tropical and temperate waters of the world (e.g., cells of Oscillatoria [Trichodesmium] erthraeus) and various types of seaweed (e.g., Sargassum, Phyllospora, Macrocystis) that detach from the seafloor and drift.

Nekton

Nekton are the active swimmers of the oceans and are often the best-known organisms of marine waters. Nekton are the top predators in most marine food chains. The distinction between nekton and plankton is not always sharp. many large marine animals, such as marlin and tuna, spend the larval stage of their lives as plankton and their adult stage as large and active members of the nekton. Other organisms such as krill are referred to as both micronekton and macrozooplankton.

The vast majority of nekton are vertebrates (e.g., fishes, reptiles, and mammals), mollusks, and crustaceans. The most numerous group of nekton are the fishes, with approximately 16,000 species. Nekton are found at all depths and latitudes of marine waters. Whales, penguins, seals, and icefish abound in polar waters. Lantern fish (family Myctophidae) are common in the aphotic zone along with gulpers (Saccopharynx), whalefish (family Cetomimidae), seven-gilled sharks, and others. Nekton diversity is greatest in tropical waters, where in particular there are large numbers of fish species.

The largest animals on the Earth, the blue whales (Balaenoptera musculus), which grow to 25 to 30 metres long, are members of the nekton. These huge mammals and other baleen whales (order Mysticeti), which are distinguished by fine filtering plates in their mouths, feed on plankton and micronekton as do whale sharks (Rhinocodon typus), the largest fish in the world (usually 12 to 14 metres long, with some reaching 17 metres). The largest carnivores that consume large prey include the toothed whales (order Odontoceti—for example, the killer whales, Orcinus orca), great white sharks (Carcharodon carcharias), tiger sharks (Galeocerdo cuvier), black marlin (Makaira indica), bluefin tuna (Thunnus thynnus), and giant groupers (Epinephelus lanceolatus).

Nekton form the basis of important fisheries around the world. Vast schools of small anchovies, herring, and sardines generally account for one-quarter to one-third of the annual harvest from the ocean. Squid are also economically valuable nekton. Halibut, sole, and cod are demersal (i.e., bottom-dwelling) fish that are commercially important as food for humans. They are generally caught in continental shelf waters. Because pelagic nekton often abound in areas of upwelling where the waters are nutrient-rich, these regions also are major fishing areas.

Benthos

Organisms are abundant in surface sediments of the continental shelf and in deeper waters, with a great diversity found in or on sediments. In shallow waters, beds of seagrass provide a rich habitat for polychaete worms, crustaceans (e.g., amphipods), and fishes. On the surface of and within intertidal sediments most animal activities are influenced strongly by the state of the tide. On many sediments in the photic zone, however, the only photosynthetic organisms are microscopic benthic diatoms.

Benthic organisms can be classified according to size. The macrobenthos are those organisms larger than 1 millimetre. Those that eat organic material in sediments are called deposit feeders (e.g., holothurians, echinoids, gastropods), those that feed on the plankton above are the suspension feeders (e.g., bivalves, ophiuroids, crinoids), and those that consume other fauna in the benthic assemblage are predators (e.g., starfish, gastropods). Organisms between 0.1 and 1 millimetre constitute the meiobenthos. These larger microbes, which include foraminiferans, turbellarians, and polychaetes, frequently dominate benthic food chains, filling the roles of nutrient recycler, decomposer, primary producer, and predator. The microbenthos are those organisms smaller than 1 millimetre; they include diatoms, bacteria, and ciliates.

Organic matter is decomposed aerobically by bacteria near the surface of the sediment where oxygen is abundant. The consumption of oxygen at this level, however, deprives deeper layers of oxygen, and marine sediments below the surface layer are anaerobic. The thickness of the oxygenated layer varies according to grain size, which determines how permeable the sediment is to oxygen and the amount of organic matter it contains. As oxygen concentration diminishes, anaerobic processes come to dominate. The transition layer between oxygen-rich and oxygen-poor layers is called the redox discontinuity layer and appears as a gray layer above the black anaerobic layers. Organisms have evolved various ways of coping with the lack of oxygen. Some anaerobes release hydrogen sulfide, ammonia, and other toxic reduced ions through metabolic processes. The thiobiota, made up primarily of microorganisms, metabolize sulfur. Most organisms that live below the redox layer, however, have to create an aerobic environment for themselves. Burrowing animals generate a respiratory current along their burrow systems to oxygenate their dwelling places; the influx of oxygen must be constantly maintained because the surrounding anoxic layer quickly depletes the burrow of oxygen. Many bivalves (e.g., Mya arenaria) extend long siphons upward into oxygenated waters near the surface so that they can respire and feed while remaining sheltered from predation deep in the sediment. Many large mollusks use a muscular "foot" to dig with, and in some cases they use it to propel themselves away from predators such as starfish. The consequent "irrigation" of burrow systems can create oxygen and nutrient fluxes that stimulate the production of benthic producers (e.g., diatoms).

Not all benthic organisms live within the sediment; certain benthic assemblages live on a rocky substrate. Various phyla of algae—Rhodophyta (red), Chlorophyta (green), and Phaeophyta (brown)—are abundant and diverse in the photic zone on rocky substrata and are important producers. In intertidal regions algae are most abundant and largest near the low-tide mark. Ephemeral algae such as Ulva, Enteromorpha, and coralline algae cover a broad range of the intertidal. The mix of algae species found in any particular locale is dependent on latitude and also varies greatly according to wave exposure and the activity of grazers. For example, Ascophyllum spores cannot attach to rock in even a gentle ocean surge; as a result this plant is largely restricted to sheltered shores. The fastest-growing plant—adding as much as 1 metre per day to its length—is the giant kelp, Macrocystis pyrifera, which is found on subtidal rocky reefs. These plants, which may exceed 30 metres in length, characterize benthic habitats on many temperate reefs. Large laminarian and fucoid algae are also common on temperate rocky reefs, along with the encrusting (e.g., Lithothamnion) or short tufting forms (e.g., Pterocladia). Many algae on rocky reefs are harvested for food, fertilizer, and pharmaceuticals. Macroalgae are relatively rare on tropical reefs where corals abound, but Sargassum and a diverse assemblage of short filamentous and tufting algae are found, especially at the reef crest. Sessile and slow-moving invertebrates are common on reefs. In the intertidal and subtidal regions herbivorous gastropods and urchins abound and can have a great influence on the distribution of algae. Barnacles are common sessile animals in the intertidal. In the subtidal regions, sponges, ascidians, urchins, and anemones are particularly common where light levels drop and current speeds are high. Sessile assemblages of animals are often rich and diverse in caves and under boulders.

Reef-building coral polyps (Scleractinia) are organisms of the phylum Cnidaria that create a calcareous substrate upon which a diverse array of organisms live. Approximately 700 species of corals are found in the Pacific and Indian oceans and belong to genera such as Porites, Acropora, and Montipora. Some of the world's most complex ecosystems are found on coral reefs. Zooxanthellae are the photosynthetic, single-celled algae that live symbiotically within the tissue of corals and help to build the solid calcium carbonate matrix of the reef. Reef-building corals are found only in waters warmer than 18° C; warm temperatures are necessary, along with high light intensity, for the coral-algae complex to secrete calcium carbonate. Many tropical islands are composed entirely of hundreds of metres of coral built atop volcanic rock.

Links between the Pelagic Environments and the Benthos

Considering the pelagic and benthic environments in isolation from each other should be done cautiously because the two are interlinked in many ways. For example, pelagic plankton are an important source of food for animals on soft or rocky bottoms. Suspension feeders such as anemones and barnacles filter living and dead particles from the surrounding water while detritus feeders graze on the accumulation of particulate material raining from the water column above. The molts of crustaceans, plankton feces, dead plankton, and marine snow all contribute to this rain of fallout from the pelagic environment to the ocean bottom. This fallout can be so intense in certain weather patterns—such as the El Niño condition—that benthic animals on soft bottoms are smothered and die. There also is variation in the rate of fallout of the plankton according to seasonal cycles of production. This variation can create seasonality in the abiotic zone where there is little or no variation in temperature or light. Plankton form marine sediments, and many types of fossilized protistan plankton, such as foraminiferans and coccoliths, are used to determine the age and origin of rocks.

Organisms of the Deep-sea Vents

Hydrothermal mussels: Galatheid crabs and shrimp grazing on the bacterial filaments that grow on the shells of the hydrothermal mussels covering the Northwest Eifuku volcano in the Mariana Arc region.

Producers were discovered in the aphotic zone when exploration of the deep sea by submarine became common in the 1970s. Deep-sea hydrothermal vents now are known to be relatively common in areas of tectonic activity (e.g., spreading ridges). The vents are a nonphotosynthetic source of organic carbon available to organisms. A diversity of deep-sea organisms including mussels, large bivalve clams, and vestimentiferan worms are supported by bacteria that oxidize sulfur (sulfide) and derive chemical energy from the reaction. These organisms are referred to as chemoautotrophic, or chemosynthetic, as opposed to photosynthetic, organisms. Many of the species in the vent fauna have developed symbiotic relationships with chemoautotrophic bacteria, and as a consequence the megafauna are principally responsible for the primary production in the vent assemblage. The situation is analogous to that found on coral reefs where individual coral polyps have symbiotic relationships with zooxanthellae. In addition to symbiotic bacteria there is a rich assemblage of free-living bacteria around vents. For example, Beggiatoas-like bacteria often form conspicuous weblike mats on any hard surface; these mats have been shown to have chemoautotrophic metabolism. Large numbers of brachyuran (e.g., Bythograea) and galatheid crabs, large sea anemones (e.g., Actinostola callasi), copepods, other plankton, and some fish—especially the eelpout Thermarces cerberus—are found in association with vents.

Patterns and Processes Influencing the Structure of Marine Assemblages

Distribution and Dispersal

The distribution patterns of marine organisms are influenced by physical and biological processes in both ecological time (tens of years) and geologic time (hundreds to millions of years). The shapes of the Earth's oceans have been influenced by plate tectonics, and as a consequence the distributions of fossil and extant marine organisms also have been affected. Vicariance theory argues that plate tectonics has a major role in determining biogeographic patterns. For example, Australia was once—90 million years ago—close to the South Pole and had few coral reefs. Since then Australia has been moving a few millimetres each year closer to the Equator. As a result of this movement and local oceanographic conditions, coral reef environments are extending ever so slowly southward. Dispersal may also have an important role in biogeographic patterns of abundance. The importance of dispersal varies greatly with local oceanographic features, such as the direction and intensity of currents and the biology of the organisms. Humans can also have an impact on patterns of distribution and the extinction of marine organisms. For example, fishing intensity in the Irish Sea was based on catch limits set for cod with no regard for the biology of other species. One consequence of this practice was that the local skate, which had a slow reproductive rate, was quickly fished to extinction.

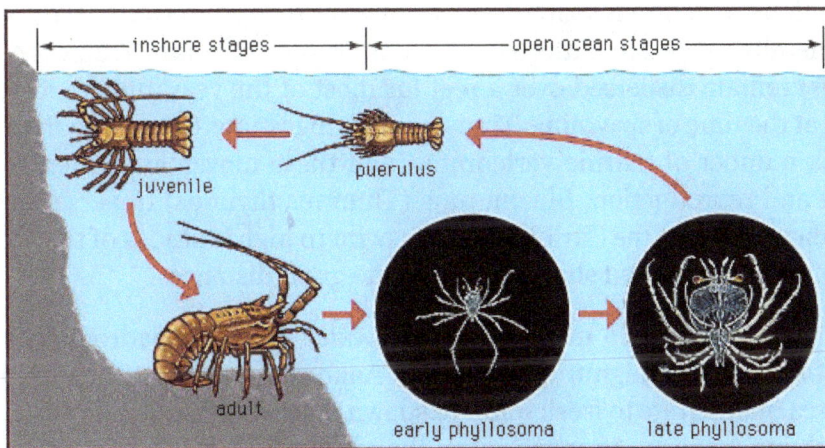

Life cycle of a palinurid lobster.

A characteristic of many marine organisms is a bipartite life cycle, which can affect the dispersal of an organism. Most animals found on soft and hard substrata, such as lobsters, crabs, barnacles, fish, polychaete worms, and sea urchins, spend their larval phase in the plankton and in this phase are dispersed most widely. The length of the larval phase, which can vary from a few minutes to hundreds of days, has a major influence on dispersal. For example, wrasses of the genus Thalassoma have a long larval life, compared with many other types of reef fish, and populations of these fish are well dispersed to the reefs of isolated volcanic islands around the Pacific. The bipartite life cycle of algae also affects their dispersal, which occurs through algal spores. Although in general, spores disperse only a short distance from adult plants, limited swimming abilities—Macrocystis spores have flagella—and storms can disperse spores over greater distances.

Migrations of Marine Organisms

The migrations of plankton and nekton throughout the water column in many parts of the world are well described. Diurnal vertical migrations are common. For example, some types of plankton, fish, and squid remain beneath the photic zone during the day, moving toward the surface after dusk and returning to the depths before dawn. It is generally argued that marine organisms migrate in response to light levels. This behaviour may be advantageous because by spending the daylight hours in the dim light or darkness beneath the photic zone plankton can avoid predators that locate their prey visually. After the Sun has set, plankton can rise to the surface waters where food is more abundant and where they can feed safely under the cover of darkness.

Larval forms can facilitate their horizontal transport along different currents by migrating vertically. This is possible because currents can differ in direction according to depth (e.g., above and below haloclines and thermoclines), as is the case in estuaries.

In coastal waters many larger invertebrates (e.g., mysids, amphipods, and polychaete worms) leave the cover of algae and sediments to migrate into the water column at night. It is thought that these animals disperse to different habitats or find mates by swimming when visual predators find it hard to see them. In some cases only one sex will emerge at night, and often that sex is morphologically better suited for swimming.

Horizontal migrations of fish that span distances of hundreds of metres to tens of kilometres are common and generally related to patterns of feeding or reproduction. Tropical coral trout (Plectropomus species) remain dispersed over a reef for most of the year, but adults will aggregate at certain locations at the time of spawning. Transoceanic migrations (greater than 1,000 kilometres) are observed in a number of marine vertebrates, and these movements often relate to requirements of feeding and reproduction. Bluefin tuna (Thunnus thynnus) traverse the Atlantic Ocean in a single year; they spawn in the Caribbean, then swim to high latitudes of the Atlantic to feed on the rich supply of fish. Turtles and sharks also migrate great distances.

Fish that spend their lives in both marine and freshwater systems (diadromous animals) exhibit some of the most spectacular migratory behaviour. Anadromous fishes (those that spend most of their lives in the sea but migrate to fresh water to spawn) such as Atlantic salmon (Salmo salar) also have unique migratory patterns. After spawning, the adults die. Newly hatched fish (alevin) emerge from spawned eggs and develop into young fry that move down rivers toward the sea. Juveniles (parr) grow into larger fish (smolt) that convene near the ocean. When the adult fish are ready to spawn, they return to the river in which they were born (natal river), using a variety of environmental cues, including the Earth's magnetic field, the Sun, and water chemistry. It is believed that the thyroid gland has a role in imprinting the water chemistry of the natal river on the fish. Freshwater eels such as the European eel (Anguilla anguilla) undertake great migrations from fresh water to spawn in the marine waters of the Sargasso Sea (catadromous migrations), where they die. Eel larvae, called leptocephalus larvae, drift back to Europe in the Gulf Stream.

Dynamics of Populations and Assemblages

A wide variety of processes influence the dynamics of marine populations of individual species and the composition of assemblages (e.g., collections of populations of different species that live in the same area). With the exception of marine mammals such as whales, fish that bear live young (e.g., embiotocid fish), and brooders (i.e., fauna that incubate their offspring until they emerge as larvae or juveniles), most marine organisms produce a large number of offspring of which few survive. Processes that affect the plankton can have a great influence on the numbers of young that survive to be recruited, or relocated, into adult populations. The survival of larvae may depend on the abundance of food at various times and in various places, the number of predators, and oceanographic features that retain larvae near suitable nursery areas. The number of organisms recruited to benthic and pelagic systems may ultimately determine the size of adult populations and therefore the relative abundance of species in marine assemblages. However, many processes can affect the survival of organisms after recruitment. Predators eat recruits, and mortality rates in prey species can vary with time and space, thus changing original population patterns established in recruitment.

Patterns of colonization and succession can have a significant impact on benthic assemblages. For example, when intertidal reefs are cleared experimentally, the assemblage of organisms that colonize the bare space often reflects the types of larvae available in local waters at the time. Tube worms may dominate if they establish themselves first; if they fail to do so, algal spores may colonize the shore first and inhibit the settlement of these worms. Competition between organisms may also play a role. Long-term data gathered over periods of more than 25 years from coral reefs have demonstrated that some corals (e.g., Acropora cytherea) competitively overgrow neighbouring corals. Physical disturbance from hurricanes destroys many corals, and during regrowth competitively inferior species can coexist with normally dominant species on the reef. Chemical defenses of sessile organisms also can deter the growth or cause increased mortality of organisms that settle on them. Ascidian larvae (e.g., Podoclavella) often avoid settling on sponges (e.g., Mycale); when this does occur, the larvae rarely reach adulthood.

Although the processes that determine species assemblages may be understood, variations occur in the composition of the plankton that make it difficult to predict patterns of colonization with great accuracy.

Shoreline Ecosystems

Shoreline and shore ecosystems happen where water meets land. Considering that water covers 75 percent of the planet, this area might appear extensive, but in reality, it comprises a narrow space. Despite this fact, much life occurs around shorelines, and the ecosystems that develop there teem with biodiversity.

Shorelines can be freshwater, saltwater or -- where rivers meet the ocean -- a mix of the two, which is called brackish water. Let's look a bit closer at some shoreline facts and about the ecosystems that exist there.

Ocean Shoreline Ecosystem

Perhaps the shoreline we're all most familiar with is the ocean shoreline we see at the beach. These ecosystems depend on the cycle of tides from high to low. Tidal pools are common in these ecosystems, which allows many aquatic animals to form special niche communities. Birds like seagulls are also common as they hunt the fish in the shallow water. Shellfish and mollusks are also found in this ecosystem attached to rocks, docks, marinas and boats.

Freshwater Shore Ecosystem

A freshwater shoreline, like the area directly surrounding a lake or a river, encompasses the shallow area near shore as well as the area on land adjoining the water. Plants form the basis of the ecosystem, and in the water, emergent plants dominate. Examples include water lilies, sedges, and arrow arum. These plants provide shelter and food to many different insects and small fish and also fertile hunting grounds for larger predators such as bass, pike, snapping turtles and wading birds.

On the shore, willows and other water-loving trees grow and provide shelter and nesting places for birds. Raccoons and other opportunist omnivores feed in the shallow water, consuming crustaceans, fish, mollusks, frogs and toads, and other shoreline animals and plants.

Estuary Ecosystems

An estuary in an ecosystem and area where saltwater and freshwater mix in one area. These are often where the mouths of rivers meet ocean environments.

The ocean exerts a strong influence in brackish river estuary ecosystems. Estuaries run by the rhythm of the tides: when the tide comes in, the water will run upstream, and when it goes out, the water will run downstream.

Salt marshes, the main type of shoreline ecosystem in estuaries, serve as nurseries of the ocean and have some of the highest levels of biodiversity in the world. Salt-tolerant grasses such as cord grass form the basis of the ecosystem. They die in the winter and provide food for a multitude of saltwater and freshwater animals.

Dune Ecosystem

Sand dunes, one of the most common shoreline types, skirt the edges of oceans and large lakes in many locations around the world. Dunes form when wind blows sand inland, where plants such as beach grass or sea grape trap the sand and it begins to pile up, creating a hill or sand dune. Although dunes may look relatively empty, many species of plants and animals inhabit them.

Insects thrive in the dry grasses where birds and spadefoot toads prey upon them. Shorebirds such as plovers and killdeer nest in low dunes. Because of high winds and tides, dunes are not permanent structures but constantly shift, move and change shape.

Shoreline Habitats

Shoreline habitats are unique areas greatly influenced by tidal patterns, neighboring estuaries and wetlands, and human uses. Each shoreline habitat supports a great diversity of life. While many of the habitats thrive, some of them require our attention so that we can monitor our human impact, and not negatively affect the abundance of life in these areas.

Rocky Shore Ecosystems

Rocky shore ecosystems are coastal shores made from solid rock. They are a tough habitat to live on yet they are home for a number of different animals and algae.

Rocky shore ecosystems are governed by the tidal movement of water. The tides create a gradient of environmental conditions moving from a terrestrial (land) to a marine ecosystem.

Ecosystems on rocky shores have bands of different species across the intertidal zone. The distribution of different species across the rocky shore is influenced by biotic and abiotic factors from above high tide to the sub-tidal zone. Different species are adapted to different environmental conditions. Some organisms can withstand being exposed to the sun for most of the day and live in the upper parts of the rocky shore. Other organisms need to be covered by the tide for most of the day and are only found lower on the rocky shore.

Zonation

Each region on the coast has a specific group of organisms that form distinct horizontal bands or zones on the rocks. The appearance of dominant species in these zones is called vertical zonation. It is a nearly universal feature of the intertidal zone.

Supratidal Zone

When the tide retreats, the upper regions become exposed to air. The organisms that live in this region are facing problems like gas exchange, desiccation, temperature changes and feeding. This upper region is called the supratidal or splash zone. It is only covered during storms and extremely high tides and is moistened by the spray of the breaking waves. Organisms are exposed to the drying heat of the sun in the summer and to extreme low temperatures in the winter. Because of these severe conditions, only a few resistant organisms live here. Common organisms are lichens. They are composed of fungi and microscopic algae living together and sharing food and energy to grow. The fungi trap moisture for both themselves and their algal symbiont. The algae on the other hand produce nutrients by photosynthesis. Green algae and cyanobacteria can also be found on the rocks of the North Atlantic coasts. They are capable of surviving on the moisture of the sea spray from waves. During the winter, they are found lower on the intertidal rocks. The algae growing higher on the rocks gradually die when the air temperature changes. At the lower edge of the splash zone, rough snails (periwinkles) graze on various types of algae. These snails are well adapted to life out of the water by trapping water in their mantle cavity or hiding in cracks of rocks. Other common animals are isopods, barnacles, limpets.

Intertidal Zone

The intertidal zone or littoral zone is the shoreward fringe of the sea bed between the highest and lowest limit of the tides. The upper limit is often controlled by physiological limits on species tolerance of temperature and drying. The lower limit is often determined by the presence of predators or competing species. Because the intertidal zone is a transition zone between the land and the sea, it causes heat stress, desiccation, oxygen depletion and reduced opportunities for feeding. At low tide, marine organisms face both heat stress and desiccation stress. The degree of this water loss and heating is determined by the body size and body shape. When body size increases, the surface area decreases so the water loss is reduced. Shape has a similar effect. Long and thin organisms dry up much faster than spherical organisms. Intertidal organisms can avoid overheating by evaporative cooling combined with circulation of body fluids. Higher-intertidal organisms are better adapted to desiccation than lower-intertidal organisms, because they encounter more hours of sun. The organisms are exposed directly to the air or they are enclosed in burrows. This results in oxygen depletion, so they can't get rid of their metabolic waste. A solution for this problem is to reduce the metabolic rate.

Intertidal zonation: at low tide, the 3
typical intertidal zones can be seen.

The intertidal zone can be divided in three zones:

- High tide zone or high intertidal zone. This region is only flooded during high tides. Organisms that you can find here are anemones, barnacles, chitons, crabs, isopods, mussels, sea stars, snails.

- Middle tide zone or mid-littoral zone. This is a turbulent zone that is (un)covered twice a day. The zone extends from the upper limit of the barnacles to the lower limit of large brown algae (e.g. Laminariales, Fucoidales). Common organisms are snails, sponges, sea stars, barnacles, mussels, sea palms, crabs.

- Low intertidal zone or lower littoral zone. This region is usually covered with water. It is only uncovered when the tide is extremely low. In contrast to the other zones, the organisms are not well adapted to long periods of dryness or to extreme temperatures. The common organisms in this region are brown seaweed, crabs, hydroids, mussels, sea cucumber, sea lettuce, sea urchins, shrimps, snails, tube worms.

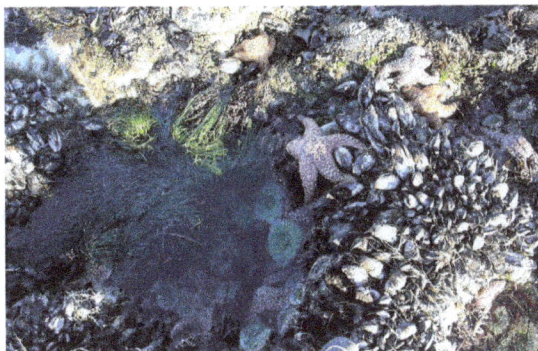

Tidal pool in Santa Cruz.

Tidal pools are rocky pools in the intertidal zone that are filled with seawater. They are formed by abrasion and weathering of less resistant rock and scouring of fractures and joints in the shore platform. This leaves holes or depressions in where seawater can be collected at high tide. They can be small and shallow or deep. The smallest ones are usually found at the high intertidal zone, whereas the bigger ones are found in the lower intertidal zone. When the tide retreats, the pool becomes isolated. Because of the regular tides, the pool is not stagnant and new water regularly enters the pool. This is necessary to avoid temperature stress, salinity stress, nutrient stress, Pools that are located higher on the beach are not regularly renewed by tides. These pools are basically freshwater or brackish water communities. It has different characteristics in comparison with other coastal habitats. Several taxa are more abundant in pools than the surrounding environment. These taxa are members of the algae and gastropods. There is also a difference between high and low located pools for the composition. In low located pools, whelks, mussels, sea urchins and Littorina littorea are common. Periwinkles and Littorina rudis are found in high located pools. Other organisms that are commonly found in pools are flatworms, rotifers, cladocerans, copepods, ostracods, barnacles, amphipods, isopods, chironomid larvae and oligochaetes. Vertical zonation also has been documented in tidal pools.

Subtidal Zone

The subtidal zone or sublittoral zone is the region below the intertidal zone and is continuously covered by water. This zone is much more stable than the intertidal zone. Temperature, water pressure and sunlight radiation remain nearly constant. Organisms do not dry out as often as organisms higher on the beach. They grow much faster and are better in competition for the same niche. More essential nutrients are acquired from the water and they are buffered from extreme changes in temperature.

Problems and Adaptations

In this topic, the problems and the adaptations are discussed. The continuously changing environment makes that organisms have to be tolerant for these changes. Adaptations are a solution for these problems and are necessary to survive.

Air

Intertidal organisms are regularly exposed to air and water. Air differs physically from seawater in diverse and important features. This influences the ability to exchange gas and their overall thermal balance with the surrounding environment. Under water, organisms are generally buoyant, because of their lower density. In air, gravity induces retraction of tentacles and other feeding organs. It also makes the body less resistant. For this reason, organisms need supporting structures when they are exposed to air. Attachment and body changes are also required. When exposed to the air, organisms directly absorb solar radiation. The buffering capacity of water, because of the high rate of heat conductivity, disappears and the body temperature increases. In contrast to this, heat loss is much lower in air than in water. An adaptation to heating is the vaporization of internal water reserves.

Light

Sunlight is another parameter that influences the organisms. When there is too much sunlight, organisms dry out and the capacity to capture light energy can be weakened. The light that is not used or dissipated can cause damage to subcellular structures. Too little sunlight reduces the growth and reproduction of the organism, because photosynthesis is reduced. Algae can avoid absorbing too much light by changing the complement or amount of pigments they produce. They also can rearrange the pigmented organelles within their cells. When free radicals are produced from an excess of light, they can be scavenged and deactivated.

Temperature

The intertidal zone can experience extreme temperature changes. The organisms in this zone must be resistant to these changes to survive. Most of the marine organisms are ectothermic and need the warmth from the environment to survive. When the organisms are submerged, they are buffered against temperature changes, because the water is isothermal. When the organisms become exposed to the air, they can experience cool or warm temperatures. When the temperature is too low, the organisms must cope with physiological threats associated with cold stress. This can be the case in polar and temperate latitude coastal zones. The body fluids can then reach their freezing point and ice crystals develop. This causes damage to cell membranes and increasing the osmotic concentration of the remaining fluids. To avoid this cold stress, organisms can migrate to habitats that are more suitable. This can be a problem for sessile organisms. They can develop physiological and behavioral adaptations such as gaping shells (mussels). Some organisms have developed antifreeze proteins. Increasing the concentration of small osmolytes such as glycerol in the body fluids can decrease the freezing point. Another strategy is to control ice crystal formation. Organisms can control the speed and the exact location of the ice crystals. When the ice formation is intracellular, it is lethal but extracellular ice formation can be tolerated. When the temperature is too high, heat stress appears. Heat stress accelerates rates of metabolic processes. This can be avoided by evaporative cooling combined with circulation of body fluids.

Salinity Stress

Salinity stress can occur in the external medium and in surface films. The concentration of the fluids determines whether or not the organism will lose water. When the osmolality of the cell is lower than the surrounding medium, the cell loses water from the internal fluids to the environment (hyperosmotic stress). When the intracellular osmolality is higher than the environment, there is an influx of water into the cell from the environment (hypoosmotic stress). Multicellular organisms respond to this salinity stress by compartmentalization. This buffers the cells from sharp changes in the osmotic environment. When the tissue has an immediate contact with the external medium, a solution can be to regulate intercellular osmotic pressure by actively excreting salts or water. Another solution is to change the internal osmolality. This can be done by incorporating ions or compatible solutes in the internal fluids.

Desiccation Stress

Organisms are threatened by desiccation during emersion at low tides or when they are positioned in the high intertidal zones. Deshydratation due to evaporative water loss is the most common mechanism. Highly mobile organisms can avoid the desiccation by migrating to a region that is more suitable. Less mobile organisms restrict various activities (reduced metabolism) and attach more firmly to the substrate. Physiological features to tolerate water loss are desiccation-resistant egg cases, reduction in water permeability of membranes, accumulation of metabolic end products, reduction of metabolic and developmental rates, maintenance of intracellular osmolytes and gene expression for production of protective macromolecules.

Predation

A wide variety of strategies to escape from predation exists. The first strategy is calcification. It makes it more difficult for the predator to eat these organisms. This strategy is applied by algae. It makes them tougher and less nutritious. A second one is the production of chemicals, usually produced as secondary metabolites. These chemicals can be produced all the time such as toxins, but other chemicals are only produced in response to stimuli (inducible defence). Another way to avoid predation is to have two distinct anatomical forms within one life cycle. This can be e.g. an alternation between a crusty form when the predator is present and a more delicate form (e.g. blade) when the predator is absent. Also the shape of the body can be a distinct evolutionary advantage. Bioluminescence is another strategy to avoid predators. Many intertidal and subtidal predators visually forage. The light is used for warning, blinding, making scare, misleading or attracting the predator. A commonly used form of protection against predation is camouflage. This can be visually or chemically. Visual camouflage means that the prey becomes invisible to the predator by using the same colors as the environment. Chemical camouflage is the passive adsorption of chemicals. The predator does not smell the prey anymore, because the smell is masked.

Wave Action

One way to protect organisms from waves is permanent attachment. But this strategy cannot be used by organisms that have to move to feed themselves. These organisms have to make a compromise between mobility and attachment. Another way to be protected is to burrow themselves into the sediment. But an alternative is to seek protected habitats. Attachment can be done by different structures. Bivalves usually use threads (byssal threads) to attach to rocky surfaces or to other organisms. But it can also be done by a foot. Another one is cementation. This is the case for bivalves such as oysters, scallops and some other forms. They lay on their side, with the lower valve cemented firmly to the bottom. This can be combined by reduction or enlargement of certain muscles.

Coral Reef Ecosystems

Coral reefs are the most diverse of all marine ecosystems. They teem with life, with perhaps one-quarter of all ocean species depending on reefs for food and shelter. This is a remarkable statistic when you consider that reefs cover just a tiny fraction (less than one percent) of the earth's surface and less than two percent of the ocean bottom. Because they are so diverse, coral reefs are often called the rainforests of the sea.

Coral reefs are also very important to people. The value of coral reefs has been estimated at 30 billion U.S. dollars and perhaps as much as 172 billion U.S. dollars each year, providing food, protection of shorelines, jobs based on tourism, and even medicines.

Unfortunately, people also pose the greatest threat to coral reefs. Overfishing and destructive fishing, pollution, warming, changing ocean chemistry, and invasive species are all taking a huge toll. In some places, reefs have been entirely destroyed, and in many places reefs today are a pale shadow of what they once were.

Corals

Animal, Vegetable and Mineral

The brownish-green specks are the zooxanthellae that most
shallow, warm-water corals depend on for much of their food.

Corals are related to sea anemones, and they all share the same simple structure, the polyp. The polyp is like a tin can open at just one end: the open end has a mouth surrounded by a ring of tentacles. The tentacles have stinging cells, called nematocysts that allow the coral polyp to capture small organisms that swim too close. Inside the body of the polyp are digestive and reproductive tissues. Corals differ from sea anemones in their production of a mineral skeleton.

Shallow water corals that live in warm water often have another source of food, the zooxanthellae. These single-celled algae photosynthesize and pass some of the food they make from the sun's energy to their hosts, and in exchange the coral animal gives nutrients to the algae. It is this relationship that allows shallow water corals to grow fast enough to build the enormous structures we call reefs. The zooxanthellae also provide much of the green, brown, and reddish colors that corals have. The less common purple, blue, and mauve colors found in some corals the coral makes itself.

Coral Diversity

Flower-like clusters of pink polyps make up this coral colony.

In the so-called true stony corals, which compose most tropical reefs, each polyp sits in a cup made of calcium carbonate. Stony corals are the most important reef builders, but organpipe corals, precious red corals, and blue corals also have stony skeletons. There are also corals that use more flexible materials or tiny stiff rods to build their skeletons—the seafans and sea rods, the rubbery soft corals, and the black corals.

The family tree of the animals we call corals is complicated, and some groups are more closely related to each other than are others. All but the fire corals (named for their strong sting) are anthozoans, which are divided into two main groups. The hexacorals (including the true stony corals and black corals, as well as the sea anemones) have smooth tentacles, often in multiples of six, and the octocorals (soft corals, seafans, organpipe corals and blue corals) have eight tentacles, each of which has tiny branches running along the sides. All corals are in the phylum Cnidaria, the same as jellyfish.

Reproduction

A purple hard coral releases bundles of pink
eggs glued together with sperm.

Corals have multiple reproductive strategies – they can be male or female or both, and can reproduce either asexually or sexually. Asexual reproduction is important for increasing the size of the colony, and sexual reproduction increases genetic diversity and starts new colonies that can be far from the parents.

Asexual Reproduction

Asexual reproduction results in polyps or colonies that are clones of each other - this can occur through either budding or fragmentation. Budding is when a coral polyp reaches a certain size and divides, producing a genetically identical new polyp. Corals do this throughout their lifetime. Sometimes a part of a colony breaks off and forms a new colony. This is called fragmentation, which can occur as a result of a disturbance such as a storm or being hit by fishing equipment.

Sexual Reproduction

In sexual reproduction, eggs are fertilized by sperm, usually from another colony, and develop into a free-swimming larva. There are two types of sexual reproduction in corals, external and internal. Depending on the species and type of fertilization, the larvae settle on a suitable substrate and become polyps after a few days or weeks, although some can settle within a few hours.

Most stony corals are broadcast spawners and fertilization occurs outside the body (external fertilization). Colonies release huge numbers of eggs and sperm that are often glued into bundles (one bundle per polyp) that float towards the surface. Spawning often occurs just once a year and in some places is synchronized for all individuals of the same species in an area. This type of mass spawning usually occurs at night and is quite a spectacle. Some corals brood their eggs in the body of the polyp and release sperm into the water. As the sperm sink, polyps containing eggs take them in and fertilization occurs inside the body (internal fertilization). Brooders often reproduce several times a year on a lunar cycle.

From Corals to Reefs

Coral Growth

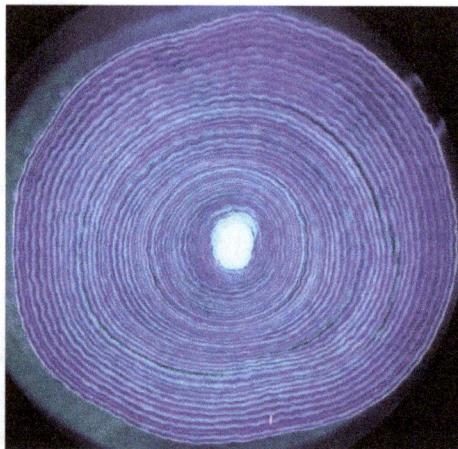

Ultraviolet light illuminates growth rings in a cross-section
of 44-year-old Primnoa resedaeformis coral found about
400 m (1,312 ft) deep off the coast of Newfoundland.

Individual coral polyps within a reef are typically very small—usually less than half an inch (or ~1.5 cm) in diameter. The largest polyps are found in mushroom corals, which can be more than 5 inches across. But because corals are colonial, the size of a colony can be much larger: big mounds can be the size of a small car, and a single branching colony can cover an entire reef.

Reefs, which are usually made up of many colonies, are much bigger still. The largest coral reef is the Great Barrier Reef, which spans 1,600 miles (2,600 km) off the east coast of Australia. It is so large that it can be seen from space.

Reefs form when corals grow in shallow water close to the shore of continents or smaller islands. The majority of coral reefs are called fringe reefs because they fringe the coastline of a nearby landmass. But when a coral reef grows around a volcanic island something interesting occurs. Over millions of years, the volcano gradually sinks, as the corals continue to grow, both upward towards the surface and out towards the open ocean. Over time, a lagoon forms between the corals and the sinking island and a barrier reef forms around the lagoon. Eventually, the volcano is completely submerged and only the ring of corals remains. This is called an atoll. Waves may eventually pile sand and coral debris on top of the growing corals in the atoll, creating a strip of land. Many of the Marshall Islands, a system of islands in the Pacific Ocean and home to the Marshallese, are atolls.

It takes a long time to grow a big coral colony or a coral reef, because each coral grows slowly. The fastest corals expand at more than 6 inches (15 cm) per year, but most grow less than an inch per year. Reefs themselves grow even more slowly because after the corals die, they break into smaller pieces and become compacted. Individual colonies can often live decades to centuries, and some deep-sea colonies have lived more than 4000 years. One way we know this is because corals lay down annual rings, just as trees do. These skeletons can tell us about what conditions were like hundreds or thousands of years ago. The Great Barrier Reef as it exists today began growing about 20,000 years ago.

Occurrence of Reefs

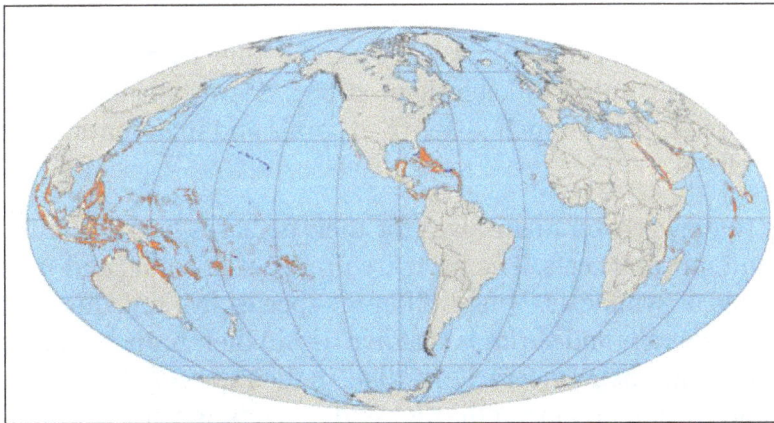

Shallow water coral reefs straddle the equator worldwide.

Corals are found across the world's ocean, in both shallow and deep water, but reef-building corals are only found in shallow tropical and subtropical waters. This is because the algae found in their tissues need light for photosynthesis and they prefer water temperatures between 70-85 °F (22-29 °C).

There are also deep-sea corals that thrive in cold, dark water at depths of up to 20,000 feet (6,000 m). Both stony corals and soft corals can be found in the deep sea. Deep-sea corals do not have the same algae and do not need sunlight or warm water to survive, but they also grow very slowly. One place to find them is on underwater peaks called seamounts.

Reefs as Ecosystems

Cities of the Sea

Scientists have been studying why populations of crown-of-thorns sea stars (Acanthaster planci) have mushroomed in recent decades. Coral reefs can suffer when the sea star's numbers explode; the echinoderm has a healthy appetite and few predators.

Reefs are the big cities of the sea. They exist because the growth of corals matches or exceeds the death of corals – think of it as a race between the construction cranes (new coral skeleton) and the wrecking balls (the organisms that kill coral and chew their skeletons into sand).

When corals are babies floating in the plankton, they can be eaten by many animals. They are less tasty once they settle down and secrete a skeleton, but some fish, worms, snails and sea stars prey on adult corals. Crown-of-thorns sea stars are particularly voracious predators in many parts of the Pacific Ocean. Population explosions of these predators can result in a reef being covered with tens of thousands of these starfish, with most of the coral killed in less than a year.

Corals also have to worry about competitors. They use the same nematocysts that catch their food to sting other encroaching corals and keep them at bay. Seaweeds are a particularly dangerous competitor, as they typically grow much faster than corals and may contain nasty chemicals that injure the coral as well.

Corals do not have to only rely on themselves for their defenses because mutualisms (beneficial relationships) abound on coral reefs. The partnership between corals and their zooxanthellae is one of many examples of symbiosis, where different species live together and help each other. Some coral colonies have crabs and shrimps that live within their branches and defend their home against coral predators with their pincers. Parrotfish, in their quest to find seaweed, will often bite off chunks of coral and will later poop out the digested remains as sand. One kind of goby chews up a particularly nasty seaweed, and even benefits by becoming more poisonous itself.

Conservation

Threats

1. Global

    The greatest threats to reefs are rising water temperatures and ocean acidification linked to rising carbon dioxide levels. High water temperatures cause corals to lose the microscopic algae that produce the food corals need—a condition known as coral bleaching. Severe or prolonged bleaching can kill coral colonies or leave them vulnerable to other threats. Meanwhile, ocean acidification means more acidic seawater, which makes it more difficult for corals to build their calcium carbonate skeletons. And if acidification gets severe enough, it could even break apart the existing skeletons that already provide the structure for reefs. Scientists predict that by 2085 ocean conditions will be acidic enough for corals around the globe to begin to dissolve. For one reef in Hawaii this is already a reality.

These bleached corals in the Gulf of Mexico are the result of increased water temperatures.

2. Local

Unfortunately, warming and more acid seas are not the only threats to coral reefs. Overfishing and overharvesting of corals also disrupt reef ecosystems. If care is not taken, boat anchors and divers can scar reefs. Invasive species can also threaten coral reefs. The lionfish, native to Indo-Pacific waters, has a fast-growing population in waters of the Atlantic Ocean. With such large numbers the fish could greatly impact coral reef ecosystems through consumption of, and competition with, native coral reef animals.

Even activities that take place far from reefs can have an impact. Runoff from lawns, sewage, cities, and farms feeds algae that can overwhelm reefs. Deforestation hastens soil erosion, which clouds water—smothering corals.

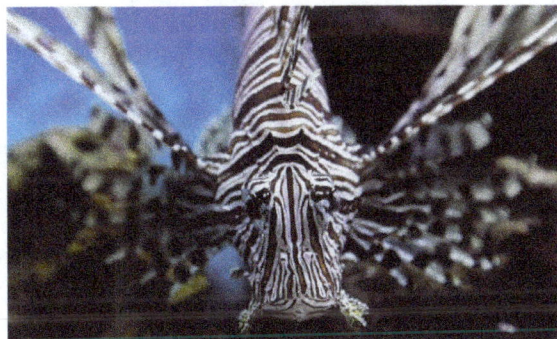

Lionfish are referred to as turkeyfish because, depending on how you view them, their spines can resemble the plumage of a turkey.

Coral Bleaching

"Coral bleaching" occurs when coral polyps lose their symbiotic algae, the zooxanthellae. Without their zooxanthellae, the living tissues are nearly transparent, and you can see right through to the stony skeleton, which is white, hence the name coral bleaching. Many different kinds of stressors can cause coral bleaching – water that is too cold or too hot, too much or too little light, or the dilution of seawater by lots of fresh water can all cause coral bleaching. The biggest cause of bleaching today has been rising temperatures caused by global warming. Temperatures more than 2 degrees F (or 1 degree C) above the normal seasonal maximimum can cause bleaching. Bleached corals do not die right away, but if temperatures are very hot or are too warm for a long time, corals either die from starvation or disease. In 1998, 80 percent of the corals in the Indian Ocean bleached and 20 percent died.

Compare the healthy coral on the left
with the bleached coral on the right.

Protecting Coral Reefs

There is much that we can do locally to protect coral reefs, by making sure there is a healthy fish community and that the water surrounding the reefs is clean. Well-protected reefs today typically have much healthier coral populations, and are more resilient (better able to recover from natural disasters such as typhoons and hurricanes).

A bluefin trevally swims in Hawaii's Maro Coral Reef, part
of the Papahānaumokuākea Marine National Monument.

Fish play important roles on coral reefs, particularly the fish that eat seaweeds and keep them from smothering corals, which grow more slowly than the seaweeds. Fish also eat the predators of corals, such as crown of thorns starfish. Marine protected areas (MPAs) are an important tool for keeping reefs healthy. Large MPAs protect the Great Barrier Reef and the Northwestern Hawaiian Islands, for example, and in June 2012, Australia created the largest marine reserve network in the world. Smaller ones, managed by local communities, have been very successful in developing countries.

Clean water is also important. Erosion on land causes rivers to dump mud on reefs, smothering and killing corals. Seawater with too many nutrients speeds up the growth of seaweeds and increases the food for predators of corals when they are developing as larvae in the plankton. Clean water depends on careful use of the land, avoiding too many fertilizers and erosion caused by deforestation and certain construction practices. In the long run, however, the future of coral reefs will depend on reducing carbon dioxide in the atmosphere, which is increasing rapidly due to burning of fossil fuels. Carbon dioxide is both warming the ocean, resulting in coral bleaching, and changing the chemistry of the ocean, causing ocean acidification. Both making it harder for corals to build their skeletons.

Open Ocean Ecosystems

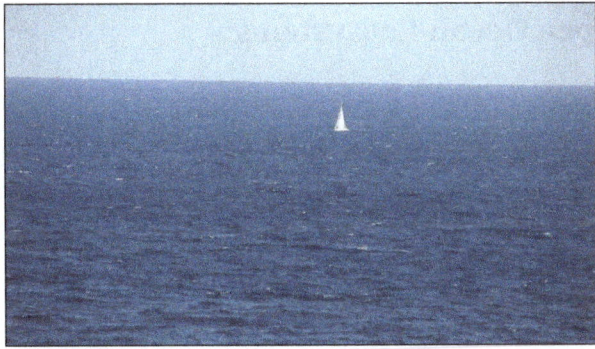

The open oceans or pelagic ecosystems are the areas away from the coastal boundaries and above the seabed. It encompasses the entire water column and lies beyond the edge of the continental shelf. It extends from the tropics to the Polar Regions and from the sea surface to the abyssal depths. It is a highly heterogeneous and dynamic habitat. Physical processes control the biological activities and lead to substantial geographic variability in production.

Zonation

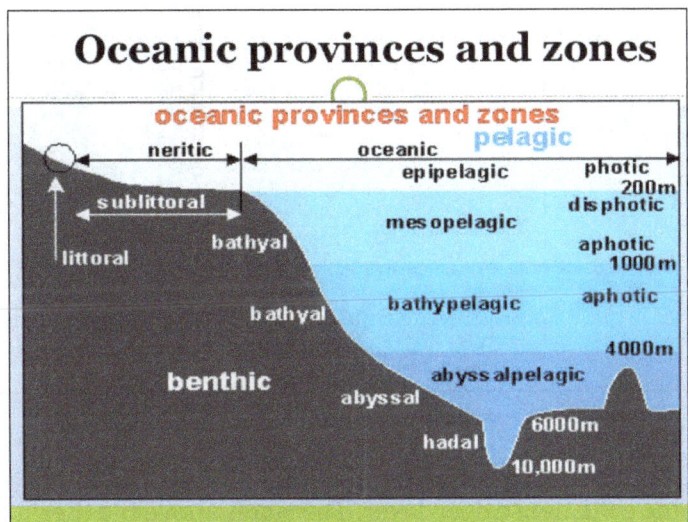

Ocean zonation.

The water column is subdivided into distinct zones by water depth and distance from shore. This is based on water depth and population composition. The distinct zones are:

- The epipelagic zone ranges from the sea surface to a depth of about 200 metres. This is also the limit of the photic zone. Light penetrates into this surface water and is usually enough for the photosynthesis.

- The mesopelagic zone lies underneath the epipelagic zone and extends to about 1,000 metres.

- The bathypelagic zone is the zone between the 1,000 and 4,000 metres.

- The abyssalpelagic zone extends to a water depth of 6,000 metres.

- The hadalpelagic zone is deeper than 6,000 metres and is found in deep-sea trenches.

Characteristics of Open Ocean Ecosystems

Currents

Currents are formed by differences in density and wind action. Density is dependent on the temperature and the salinity of the water. Cold and salty water is dense and will sink. Warmless salty water will float. There are different types of currents such as deep water circulation, surface circulation, wind-induced currents and nearshore currents.

Stratification

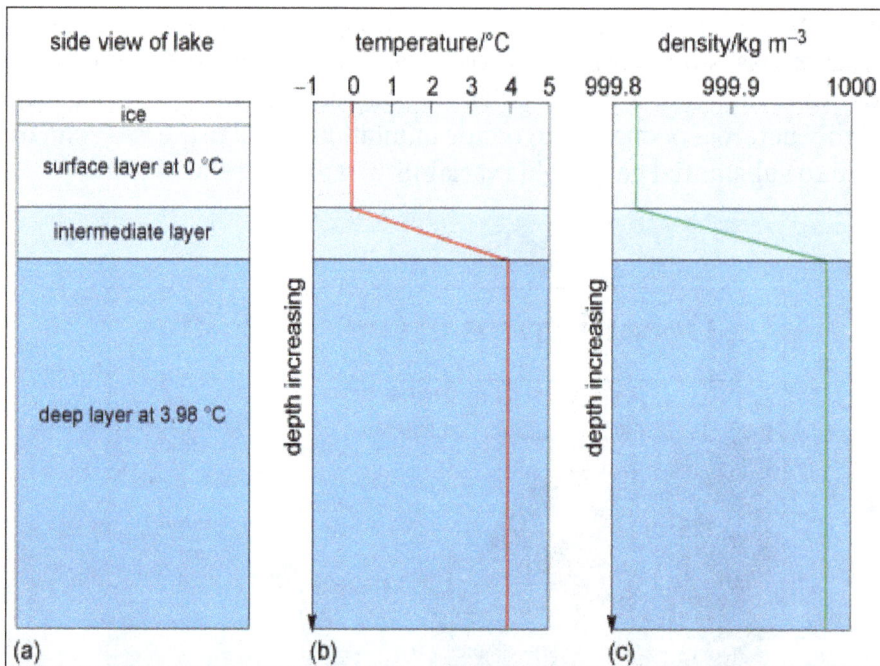

Stratification in the ocean, based on 3 physical characteristics.

The water column can be stratified by density. A permanent thermocline separates the warm, buoyant surface layer from the cold, dense deep layer. The water above the thermocline is well mixed by the wind. Also the temperature, salinity and the community of organisms in this water change with the seasons. Below the thermocline, the water is isolated from the atmosphere so the temperature and salinity remain stable over the year. This water is sparsely populated with animals. Besides the temperature, stratification can also be caused by differences is salinity and differences in density. The transition zone is called a halocline for the salinity and a pycnocline for the density. In the temperate latitudes, heat is gained and lost during the summer and winter. Together with this, a seasonal thermocline at about 100 meters depth comes and goes.

Light Transmission and Absorption

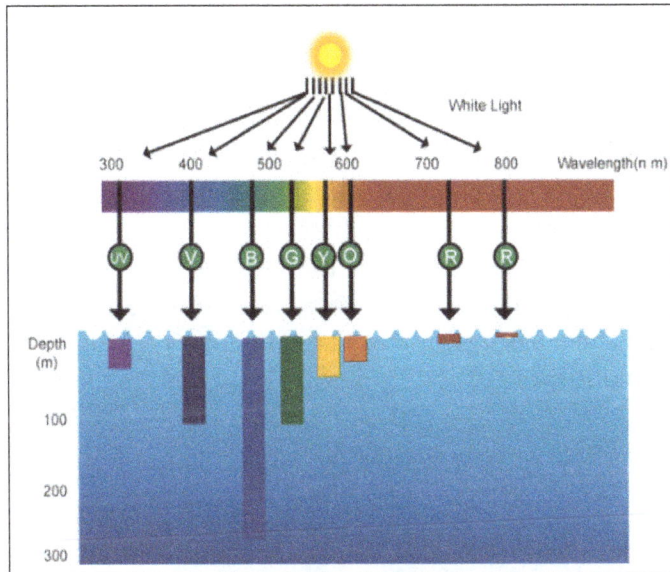

Light absorption in the open ocean.

Life depends directly or indirectly on energy from sunlight. In the ocean, marine plants and protoctists use green chlorophyll and a few accessory pigments to capture the visible light from the sun. A large fraction of that sunlight is reflected from the sea surface back to the atmosphere. The remaining light enters the water and is absorbed by water molecules. Approximately 65% of the visible light in water is absorbed within 1 meter of the sea surface. This energy is converted into heat and elevates the surface water temperature. The red and yellow light (longer wavelengths λ) are absorbed by water more readily then the green and blue light (shorter wavelength λ). This is called the selective absorption of wavelengths. It accounts for the blue colour of the open ocean. In very clear water, not even 1% of the light that enters the ocean penetrates to a depth of 100 meters. Water close to the land has high suspended solids. Because of this, light may not penetrate this turbid water deeper than 20 meters. This causes a shift to the green and yellow wavelengths, because of the reflecting wavelengths by suspended particles. Due to the characteristics of this light transmission, the water column can be divided into two distinct zones. The upper zone is called the photic zone. In this zone, plants receive adequate levels of sunlight and can photosynthesize. The zone below the photic zone is called the aphotic zone. Plants cannot survive in this dark zone.

Biology

Despite the enormous size of the open ocean, it does not support a dense population of organisms in its water or on the sea bottom. This is because it is located far from land, which is the main source of the essential nutrients that organisms need to grow. The water is mixed by currents of large circulation gyres. The biomass is limited but the diversity is remarkably high. A useful parameter for delineating depth zones and the vertical distribution of the biota in it is the degree of illumination. The upper zone or photic zone is well illuminated, so plants can photosynthesize. Depending on the clarity of the water, this zone can extend to depths of 100 to 200 meters. Seasonal succession of indigenous biota is also visible. The water is nutrient – poor, because of the absence of land and rivers. Plants cannot live attached to the bottom of the open ocean because there is too little light. The sparse plants that live here are entirely planktonic. The pelagic flora falls into two main groups: the unicellular microphytoplankton and the large free-floating macroalgae. Microphytoplankton consists of diatoms, dinoflagellates and coccolithophorids. The larger diatoms are grazed by large 'zooplankton' and these are on their turn consumed by fish. Herbivorous zooplankton consists of foraminifera and radiolarian. These are consumed by larger animals such as copepods. The floating organisms (neuston) at the surface have a blue colour due to the presence of protective pigments. These pigments are able to reflect the damaging part of the light spectrum (UV). This is necessary because the habitat is exposed to high levels of ultraviolet radiation. Organisms in the epipelagic such as crustaceans adopt the red colours as a mean of camouflage. This is possible because red light is absorbed rapidly and does not penetrate far into this zone. The red colours are effectively invisible at depth and this makes the organisms invisible to others.

Neuston Physalia physalis.

The dysphotic zone is a zone with little light perception. The lower limit lies at a depth of approximately 500 to 1,000 meters. In this layer, an oxygen minimum is present. No seasonal effects of heating and cooling are present. This makes the zone stable and unchanging over time. Abundant organisms are crustaceans (copepods, shrimps, amphipods, ostracods and prawns). These organisms have red to red-orange pigmentation. Other animals are squids and fishes. Several adaptations have evolved to this zone. The first one is the presence of large size and light-sensing ability of the eyes. Some organisms have additional organs called photophores, which produce bioluminescence. Some fish have bacterial photophores in which light is produced by the metabolic activities of symbiontic bacteria that live in dense concentrations in the photophores. Some species engage in diurnal vertical migration to get food. They swim upward to the photic zone at night to feed and descend again during the day.

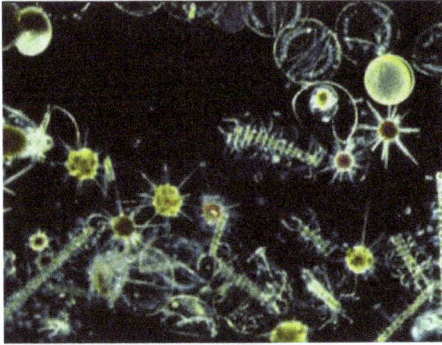
Microphytoplankton.

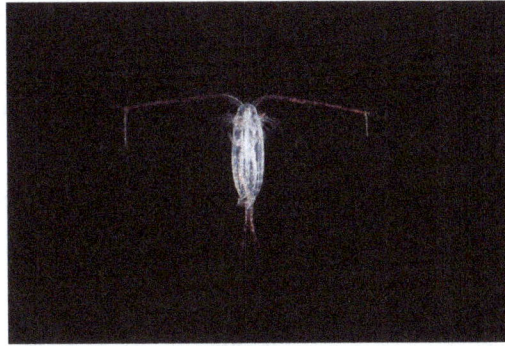
Copepod.

In the aphotic zone, it is constantly dark and cold. The zone begins at depths of 500 to 1,000 meters. Many species in this region are coloured red or black. Organisms that occur here are copepods, ostracods, jellyfishes, prawns, mysids, amphipods, worms and fishes. Fishes at mid-depth have evolved adaptations to maximize their chances of capturing their prey. These fishes are small with enormous mouths lined with sharp teeth. Their jaws can be unhinged to eat large preys. They also have bioluminescent organs. Some species have a fishing rod with an attached luminous lure in front of the head. This is the case for the female deep sea angler fish. The body musculature is reduced. The deeper zones are dominated by macrobenthos. The principle food resources on the deep-sea bottom appear to be the slow fallout of fin and coarse organic detritus from surface waters, the settling of large animal carcasses, the sinking of fecal matter and the transport of organic detritus by turbidity currents. Bacteria are also common in this region. They break down organic matter into simpler inorganic nutrients. The bacteria are consumed by heterotrophic nanoflagellates, which are in turn consumed by ciliates. This food chain is called microbial loop. This loop recycles organic matter that is too small to be consumed by metazoan plankton.

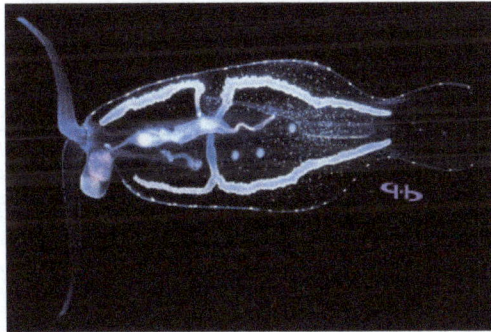
Bioluminescence in Nudibranchs.

The greater the depth, the more organisms face with increasing physiological stress. One of these stress factors is pressure. Pressure increases by 1 atmosphere for every 10 meters increase in depth. In some areas, oxygen minima layers are present. These oxygen minima are caused by the bacterial breakdown of material sinking from the sea surface.

Threats

Despite the fact that much of the open ocean is remote from the land, it has not escaped human impacts. The main problem is overfishing. To demonstrate this, 90% of stocks of large pelagic fish such as tuna and jacks are removed by fishing. The whole zooplankton communities have shifted their spatial distribution possibly in response to ocean warming. The result of this overfishing is that the fishes of higher trophic levels are replaced by fishes of lower trophic levels. This is called fishing down the marine food web. It is ecologically unsustainable. This causes the loss of many species or a degradation of the environment. The introduction of alien species is also a problem. These are species that are introduced in another area and can compete with the indigenous species. This introduction can be done through ballast water from cargo ships or on hulls of vessels.

Deep Ocean Ecosystems

The deep sea is the largest ecosystem on Earth Deep below the ocean's surface is a mysterious world that takes up 95% of Earth's living space. It could hide 20 Washington Monuments stacked on top of each other. But the deep sea remains largely unexplored. Dive down 650 feet (one monument or 200 meters), and you notice that light starts fading rapidly. Dive deeper: the temperature drops and pressure rises. At 13,000 feet (20 monuments or 4,000 meters), the temperature hovers around freezing, and there's no sunlight at all.

Deep-sea Species

The "deep-sea" water mass can be subdivided into four depth zones:

- Mesopelagic (150-1 000m);

- Bathypelagic (1 000-3 000m);

- Abyssopelagic (3 000-6 000m); and

- Hadal zone, below 6 000m depth, in the deep ocean trenches.

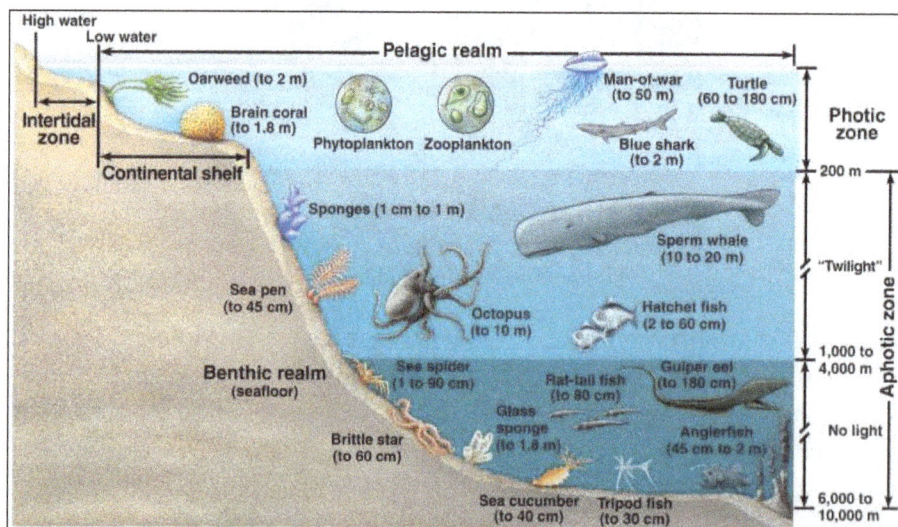

From a demersal, or seafloor perspective, the deep-sea region consists of the continental slopes (starting at the shelf break), the continental rise which extends down to the abyssal plane at around 6000m, and the trenches. The seamounts stand out of the abyssal plain.

No light penetrates beyond 1 000m and even at depths of 150m light levels are reduced to one percent of those at the surface, insufficient to support photosynthesis. Thus, organic material must be convected into the deep waters, which occurs in various ways. Dead phytoplankton and nekton sinks, and though much is consumed as it settles, sufficient amounts enter the deepwater to sustain much of the biomass there.

Many species undergo extensive diel vertical migration, a pattern of feeding in the surface waters at night and moving down during the day to reduce predation. In this way, surface production is cascaded through progressively deeper layers. Of relatively minor productive importance is organic material from large carcasses sinking to the seafloor, e.g. dead whales, and sulpha-based organic production associated with deep-sea seafloor hot-water vents. Nevertheless, the concentration of organic material decreases exponentially with depth.

In contrast to former views, it is now known that seasonal effects in surface layers are transferred into even deeper ocean regions. Therefore, despite the physical uniformity of the deep oceans, an annual production signal exists that results in seasonal migrations and reproductive cycles in deep-sea fauna.

Marine Biodiversity

Deepwater fish comprise three major groups:

- Pelagic fish living largely in midwater, with no dependence on the bottom;

- Demersal (seafloor) fish, living close to and depending on the bottom; and

- Benthopelagic fish, living close to the bottom but undertaking short migrations in the watermass (e.g. for feeding).

In general, the deep-sea demersal fish come from phylogenetically much older groups than the pelagic species (the first existing demersal species were present around 80 million years ago). While most of the demersal deep-sea families are found worldwide, the existence of isolated deepwater basins bounded by the continents and mid-oceanic ridges has resulted in regional differences believed to be a consequence of continental drift and subsequent ocean formation.

Demersal species are distributed according to depth. Those species that inhabit the continental slope and rise are spread along ribbon-like depth regions along the perimeters of the oceans. Where deepwater pelagic species and demersal species co-occur, they usually prey on each other.

In order to cope with the relatively cold and dark conditions in the deep sea, species that live there have adapted in a variety of ways. They are a diverse group of species with different life histories, productivity rates and distribution patterns. However, much remains unknown about deepwater fishes and new discoveries are continually made, such as the megamouth shark (a 4.5m and 750kg shark) and the six-gilled ray, both of which represent previously unknown families.

The importance of deep-sea species and ecosystem biodiversity has led to concerns about their increased vulnerability as a result of fishing activities. Physical vulnerabilities can occur when fishing gears come into direct contact with the seafloor or structural elements of the ecosystem.

Functional vulnerability can occur, for example, as a result of the selective removal of a species which may change the manner in which the ecosystem functions.

Some features of an ecosystem, particularly those that are physically fragile or inherently rare, may be vulnerable to most forms of disturbance, but the vulnerability of many populations, communities and habitats may vary greatly depending on the type of fishing gear used or the kind of disturbance experienced. Vulnerability is not an absolute concept: disturbances could be considered acceptable at a given time and/or location but considered to cause unacceptable damage at other times and/or locations.

Particular ecosystems are likely to show increasing vulnerability when fishing intensity increases. However, the relationship between fishing intensity and vulnerability may not be linear and proportional. In some ecosystems this relationship is more "step-like," with abrupt changes occurring once thresholds are crossed.

Ecosystem components identified as particularly vulnerable include sponge-dominated communities, coldwater corals, and seep and vent communities. These are often associated with topographical, hydrophysical or geological features such as summits and flanks of seamounts, hydrothermal vents and cold seeps.

Coryphaenoides armatus (Macrouridae) from 4800m
in the Porcupine Abyssal Plain.

Mangrove Ecosystems

Mangroves are the dominant ecosystems that line the coasts of subtropical and tropical coastlines around the world. Mangroves are survivors. With their roots submerged in water, mangrove trees thrive in hot, muddy, salty conditions that would quickly kill most plants. A series of adaptations allow these plants to not only live in these conditions but thrive, including a filtration system that keeps out much of the salt and a complex root system that holds the mangrove upright in the shifting sediments where land and water meet. Mangrove ecosystem also supports an incredible diversity of creatures, including some species unique to mangrove forests. And, as scientists are discovering, mangrove swamps are extremely important to our own well-being and to the health of the planet. There are four species of mangroves that range in size, characteristics, and preferred habitat. Mangroves act as natural buffers between the land and sea. Mangroves absorb the force of strong waves that have been built up on the open seas. This is especially important during times of tropical storms. Additionally mangroves counteract erosion which waves cause over time. From a filtration standpoint mangroves neutralize sediment runoff from both natural and human activities, mangroves are often referred to as carbon sinks. Mangroves hold vital habitat to animals like birds that nest in their tight knit canopy. More importantly aquatic organisms that we as humans rely on are found in the tangled roots of the mangroves. Many juvenile fish from the nearby coral reefs, like snapper and grouper, come to take refuge in the tangled roots of the mangroves. Here the young fishes can grow with much less competition and predation. Mangrove lagoons are an important habitat for juveniles of many fish species, and can provide nursery areas for estuarine as well as reef fishes. Crustaceans such as shrimp, crabs, and lobster also take hold and grow in these ecosystems. Some smaller species of fish never leave the tangled root systems of the mangroves. The area surrounding the edges of mangroves is the favorite hunting grounds for fish like the tarpon, rays, and lemon sharks. Another bounty for humans is the massive gastropods known as conch. Conchs are large slow moving snails that are highly sought after in the tropics of the Caribbean for their delectable meat. The mangrove wood itself has been collected for ages for fuel by those living in close proximity to the coastline. In recent years however mangroves are in decline as they are cleared for development, converted to aquiculture ponds, or wood chipped to aid in the production of rayon. This destruction displaces organisms that rely upon the mangroves for a rearing ground, hunting ground, and home. Additionally it puts its new human residents in danger of harsh waves and consequently the erosion that comes along with them.

Mangrove Species

Mangroves forests of tropical trees and shrubs rooted in saltwater sediments between the coast and the sea are crucial nurseries for coral reef fish. The tallest and most distinctive mangrove is the red mangrove (Rizophora mangle). Under optimal conditions this tree can reach 80 feet in height in some parts of the world, but on average reaches a height of 20 feet. The distinctive characteristic of this species is its reddish prop roots which can extend out of the water up to 3 feet. These roots are the primary home for the many species of organisms that thrive in the ecosystem. The red mangrove is the most tolerant of turbulent water. This mangrove is the first specie to meet the aggressive waves from the open ocean. The black mangrove (Avicennia germinans) is smaller and grows closer to the shoreline in areas slightly higher in elevation, more susceptible to tidal fluctuations. In the times of low tide pencil-like projections, or pneumatophores, extend into the atmosphere to provide oxygen for the underwater root systems. This specie can reach over 65 feet in some locations, but on average it will reach heights up to 50 feet. This tree appears similar to that of the red mangrove with opposite leaf arrangement. However the presence of pneumatophores and lack of prop roots sets them apart. The white mangrove (Laguncularia racemosa) is found at even higher elevations where soils are least saturated. The most distinguishing feature is the lack of aerial roots. Unlike the black and red mangroves the white mangrove conceals its roots underground, with the exception of prolonged flooded areas where the tree will develop a sort of peg root to obtain oxygen. The fourth specie of mangrove is associated with the upland transitional zone is restricted in its range by its extreme sensitivity to cold. The buttonwood mangrove (Conocarpus erectus) is a shrub-like tree and is least like the other mangroves but still plays a vital role in the mangrove ecosystem being home to terrestrial animals and avian nesters.

Terrestrial Inhabitants

There is a wide range of mammals that visit the mangroves to hunt and forage for food. Some small like raccoons, opossums or agoutis, but there are also large animals that rely on these ecosystems. The very rare Florida panther was last seen in the everglades mangrove system, and due to habitat destruction it is estimated that roughly 50 individuals remain in the wild. The endemic key deer is only among the mangrove systems in the Florida Keys. In many other parts of the world monkeys are very common in mangroves, and in fact have been captured on video diving in the shallows collecting food. The greater flamingo and other special birds also visits these ecosystems were they forage in the shallows for tiny crustaceans. Many waterfowl species in North America spend their wintering months in and around the mangroves searching for food before their counter migration back north. To one particular sort of bird, mangrove canopies are among one of the few options for nesting. Pelagic birds, or sea birds, are birds that spend the majority of their lives at sea. These birds are highly efficient at staying aloft and use little energy when flying. They come to land rarely and often only to nest. To these birds the thick tangled canopy of the mangroves is ideal to build for nesting and rearing young. The seeds from a parent tree drift across the sea and rest upon an exposed section of sand, taking root and virtually creating a bird paradise. Black mangrove grows in the intertidal zone throughout the Gulf of Mexico. It will establish in nature from seed that floats and can travel some distance on the tides. These specialized birds, such as the magnificent frigate bird, can nest on an island free from predation and have an open buffet on either side to feed their young.

Aquatic Inhabitants

Among the tangle of the mangrove prop roots (Rizophora mangle) lays an absolute flourish of life. Nearby are beds of seagrass which are considered part of the mangrove ecosystem. The organisms from the mangroves interact with those of the seagrass beds. In the Caribbean, juvenile reef fish occupy the submerged prop roots of Rhizophora mangle and make frequent foraging runs into adjacent seagrass beds. Mangroves serve either as nurseries for juveniles or as feeding areas for transient fish and crustaceans. For many fish they start their lives in the adjacent sea grass beds, they migrate into the mangroves as they grow slightly larger, and then when they are just about fully mature they migrate across the flats to the coral reefs. This is known as ontogenic migration. It has been found that certain coral reefs are correlated to certain mangrove systems. Therefore if a mangrove forest is removed, the fish populations of the correlating coral reef are negatively affected. Just as waterfowl seem to have a sixth sense when it comes to direction, reef fish also seem to know where their desired destination is as the set off on their final migration to the coral reef where they will live out the rest of their lives. Migrating fishes can move directly between mangroves and reef habitats, passing over deep water and seagrass beds. The reasons for undertaking such a stage-structured lifecycle are varied and include: (i) requirements for different food sources as the organism grows, (ii) changing risks of predation with size, such that sheltered habitats, where predator foraging efficiency is low, are chosen when the organism occupies its smaller, more vulnerable stages, and (iii) a need to reproduce in habitats which offer the greatest dispersal or survival of larvae.

Mangroves as a Nursery

There is a wide range of fish species that live in the nursery of the mangroves. There are some fish species that live in the mangroves fulltime, some come and go with the seasons, and some just live in the mangroves until they reach maturity. Species that live here full time are a variety of gobies, spotted trout, pipefish, and others. These full time residents never grow too large for their habitat and therefore never have reason to leave. During the rainy season, the increased flow of freshwater results in the appearance of freshwater species, such as the predatory Florida gar. This increased flush of freshwater has been known to drive some species further off shore in search of higher saline waters. Among the resident species and seasonal guests are the juvenile fish species which migrate to the coral reefs once they have reached maturity. Of these juvenile reef fish are various species of grouper, snapper, jacks, schoolmasters, and others. Among the grouper species that live in these habitats is the endangered Goliath grouper. These spectacular fish can reach weights of over 800 lbs. when mature. Goliath grouper breed in mass swarms in the open ocean and their fry grow in the seagrass beds and then they migrate to the nearby tangle of the mangroves. Goliath grouper remained in mangrove habitats for 5–6yrs (validated ages from dorsal spine sections), then emigrated from mangroves at about 1.0 m total length.

In this mangrove nursery there are many species of fishes that flourish in a predator reduced environment. The larger predatory fish are too large to pursue their prey in the mangroves. While within the mangroves themselves there are very few predators, in the more open waters are predator fish that await any small wanderers that may stray away from the protection of the mangroves. Fish like the tarpon, snook, bone fish, red fish, rays, and lemon shark work with the tides. They are fairly large fish and need deeper water to enter the shallows surrounding the mangroves. Once these predators get into the shallows they are on a race against the clock to hunt as much as they can before the tides retreat and they become stranded. These fish are opportunistic feeders taking whatever they can fit in their mouths; just about everything is fair game. These fish don't have any natural predators in these waters. However they are highly sought after by the saltwater angler.

Other Mangrove Inhabitants

The ecosystem of the mangroves is not just home to fish species. Also calling home to these parts is a wide variety of invertebrates. The extensive root systems, muddy bottoms, and open waters are all home to invertebrates that are well adapted to the temperature and salinity fluctuations as well as the tidal influences common to mangroves. These wide array of organisms feed on leaf litter, detritus, plankton, algae, and other small animals. Barnacles, mollusks, shrimp, crabs, lobsters, jellyfish, tunicates, etc. can all be found among the roots of the mangrove. There is also a very ancient animal that can be found living in the mangroves, the earliest fossils of the horseshoe crab dates back to roughly 450 million years ago.

Horseshoe crabs are scavengers and may be found among mangroves feeding on algae, invertebrates, and dead organisms. They are especially adapted to low oxygen waters, possessing up to 200 book gills used for respiration. On the edges of the mangroves and out in the seagrass beds one can find colorful anemones with their feeding tentacles sent up to catch food, sea urchins crawling about munching algae, and large gastropods known as conch scooting about the bottom. These large slow moving snails are known for their iconic shells, and have been collected for years for their meat. Also commonly found in these waters and collected for their meat is the Caribbean spiny lobster. Their habitat usage is similar to that of the juvenile reef fish. Depending on their maturity they sought different habitat type. Largely influencing their preference for habitat was predation. Predation on newly settled juveniles was greater in seagrass and coral crevices than in mangrove prop roots, whereas the survival of larger juveniles was higher in mangroves and coral patch reefs than in seagrass. These results suggest that mangrove habitats may function as a nursery for juvenile spiny lobsters but that the use of this habitat depends on shelter characteristics and the isolation of islands.

Along with invertebrates and fishes, sea turtles are also associated with mangrove vegetation, at least point in their life. The Atlantic ridley sea turtle is commonly observed in the mangroves that line the bays of south Florida. The loggerhead and green sea turtles have been known to utilize the mangroves as juvenile nurseries as well, receiving protection from predation and ample access to a food rich environment. The green sea turtle and the Hawksbill sea turtles have been observed feeding on the roots of mangroves as well as other submerged vegetation. Another reptile known to the mangroves are saltwater crocodiles. Mangroves are particularly well suited for juveniles. All around the world species of water snakes can be found living in mangroves, as well as some saltwater tolerant amphibians. The most common of these amphibians is the giant toad (bufo marinus). The giant toad, or more commonly known as the cane toad, is native to Central and South America but has been introduced in the Caribbean and some Pacific Islands. Originally introduced to control the cane beetle it is now considered a pest. There are also two quite different aquatic animals that also are very common in and around mangroves. Feeding on the small fish species dolphin are quite common to mangroves. Feeding on seagrass and submerged vegetation manatees are also often seen swimming around the edges of mangroves along the coast of Florida. Another species of manatee that can be seen in coastal marine habitats is found in West Africa.

Erosion Resistance

Besides for crucial for wildlife habitat both above and below the surface, mangroves also serve a purpose to the land beyond them. Mangroves are very efficient at counteracting erosion. In many areas around the world mangroves are removed for various reasons, and in these areas the effect of erosion can be seen almost immediately. Mangroves act as a natural wave block from the strong waves that have built up on the open ocean. Even a small band of mangrove is an efficient blocking wave. Mangroves are extremely important in times of increased wave strength, such as during tropical storms. The best example on finds is the super-cyclone which occurred on the 29th October 1999 with a wind speed of 310 km hr-1 along the Orissa coast (India) and played havoc largely in the areas devoid of mangroves. On the contrary, practically no damage occurred in regions with luxuriant mangrove growth (K Kathiresan). This example along should be evidence enough to prove that mangroves are essential in protecting coastal lands. Ironically the highly populated areas most in need of protection from tropical storms are the areas most likely to remove mangroves. A prime example of this would be Cancun, Mexico. Cancun is situated at the very tip of the Yucatan Peninsula, which is a relatively large piece of land situated in the middle of the Caribbean Sea. Given its location, the peninsula is highly susceptible to hurricane activity and with the removal of their mangroves they are rendered helpless to any strong tropical storm. The logic for the removal is for more pleasing aesthetics, yet less aesthetically pleasing are the sandbags set in place to combat the waves and high water. Without the mangroves to hold sediment in place, even the gentlest of waves will take some amount of sediment out into the ocean. A wellknown and credible institute has witnessed this first hand. The Smithsonian Institute operates a field station on a small island, 0.4 hectare in size, 14 miles off the cost of Dangriga, Belize. The Carrie Boy Field Station is located on the Meso-American Barrier Reef, and is devoid of mangroves. Do to this absence of mangroves the already small island shrinks every year, at an average decreasing rate of 2 inches. Along with keeping sediment in place, mangroves also filter the water passing through the sediment. "Black mangrove is valuable in restoring brackish and salt water marshes due to its ability to filter and trap sediments. Mangrove forests, which include black mangrove, have a high capacity as a sink for excess nutrients and pollutants. It also mixes well with other native plants to reduce wave energy". Mangroves are so efficient at filtering nutrients from water that in some places they are being used on a municipal level to treat sewage. A mangrove wetland was constructed in Futian, Shenzhen, China to do just this. Three belts of mangroves were planted, housing three different species. The belts through which the water was fed held dimensions of 33 m in length, 3m in width, and 0.5 m in depth. Expectedly the removal of organic matter and nutrients were positively correlated with plant growth. The results indicated that mangroves could be used in a constructed wetland for municipal sewage treatment, providing post-treatment to remove coliforms was also included. In the natural setting filtration is key as well, especially in areas where development is prevalent. The filtration of runoff is particularly important where the surrounding region contains seagrass or coral reefs. These habitats are particularly vulnerable to deterioration if water quality declines. If coral reefs and seagrass habitats were to be lost, numerous highly valuable ecosystem goods and services would also be lost.

Economic Importance

Mangroves are economically important as well. The value of mangrove ecosystem services world-wide has been estimated as an annual global flow of US$ 1,648 billion. The problem is when harvest exceeds production. Of 2500 households in 21 rural communities (about 13.000 people) near the Caeté estuary, 83% derive subsistence income, and 68% cash income through use of mangrove resources. Near this estuary on the northern Brazilian coast, everything is used in the ecosystem from the sea, to the land, and even the wood itself. The mangrove crab (Ucides cordatus) is collected and sold by 42% of households, and constitutes a main income source for 38%. Including processing and trading occupations, over half of the investigated populations depend on the mangrove crab for financial income. In addition to the mangrove crab in these regions, many households depend on the use and sale of the mangrove wood and bark. Not only in Brazil but virtually worldwide mangrove wood is collected on both the small scale and commercial scale. Mangrove wood is high in tannin therefore is very sought after for its durability in the timber industry. As with any wood harvest it is done both sustainably and not sustainably. In fact, mangroves can be very successfully managed for sustainable timber production – the Matang Mangroves in western Peninsular Malaysia are a classic example. The management system, first started at the beginning of the 20th century, is based on a 30 year rotation. Small areas are clear-felled in a patchwork manner, and allowed to regrow before the next harvest. Mangrove wood burns with a high heat output and no smoke, making it ideal fuel wood. Coastal rural inhabitants have been using mangrove as fuel for many generations. Consequently these characteristics make it very attractive for mass collection coal production. One ton of mangrove wood is equivalent to 5 tons of Indian coal. Due to its durability the rayon industry also ravages the mangroves for their wood, as does the wood chip industry. These three industries are highly responsible for the destruction of mangroves. The pneumatophores can be used to make bottle stoppers and floats for fishing. A perhaps unexpected industry is the mangrove honey industry. During full bloom bees and other insects are found buzzing about mangroves as they are any patch of wild flowers. And in doing so collect pollen on their bodies which is later brought to the hive and put to use. Many people in foreign lands know of the sweet nectar produced and harvest this honey produced. For instance, the Sundarbans provide employment to 2000 people engaged in extracting 111 tons of honey annually and this accounts for about 90% of honey production among the mangroves of India. In Bangladesh, an estimated 185 tons of honey and 44.4 tons of wax are harvested each year in the western part of the mangrove forest.

Fisheries

Perhaps the most exploited, but economically profitable, parts of the mangrove ecosystem are the fisheries. On a subsistence level, fishing can be done sustainably and is quite crucial for some populations of people. On a local level many individuals dive for the Conch and keep themselves as well as sell them to local restaurants. But on a commercial level, animals are taken out of the mangroves for money and often at a rate at which the ecosystem cannot supply for the demand. Most common are shrimp and the various species of fish, although other crustaceans and shellfish are also sought after. Implementing the proper techniques the fisheries can be harvested from on a sustainable level. Most of the time the fishermen are often overwhelmed by greed, which ultimately leads to the down fall for the fishery. And when the ecosystem does not provide enough for what the demand is, aquaculture is established. Typically this means that the natural mangrove ecosystem is destroyed, and what remains is manipulated to better suit for a certain organism in order to gain cash profit. Consequently this only suits those particular organisms banishing the others from existing there. The food and other chemicals put into the water also adversely affect the rest of the surrounding waters around these man made ponds. The driving force is the high demand for shrimps in the developed countries and the governments of developing countries grossly undervaluing their mangroves.

Carbon Sequestration

Beyond the economical possibilities and the crucial wildlife habitat for so many varieties of organisms, is yet an even more important value of the mangrove. This is one that is important not only in small communities or even whole countries, but on a global scale. This unique characteristic is the mangroves ability to remove carbon from the atmosphere. Much like the wetlands of the inlands and the boreal forests of the temperate climates, mangrove systems are referred to as carbon sinks. The ocean itself does a great job through its direct airto-sea exchange activity. This more or less is the action of waves capturing air and bringing the carbon into the ocean. With the combined oceanic air-to-sea exchange and the mangroves ability to seize atmospheric carbon, the two seize a huge amount of carbon. Measurements show that mangroves have the ability to seize around one and half tons of carbon per hectare per year. This is approximately equivalent to the amount of carbon a motor vehicle releases to the atmosphere each year (assuming each car uses approximately 2,500 liters of petrol per year). Indonesia has 4.5 million hectares of mangrove, and from this previous information we can conclude that the emission from 5 million cars can be erased by the mangrove ecosystem in Indonesian alone. A high amount of carbon is stored in the upper layer of mangrove sediment, around 10 %. Each hectare of mangrove sediment consequently contains about 700 tons of carbon per every meter of depth, and digging up the sediment to create ponds for aquaculture results in the potential oxidation of 1/400 tons per hectare per year. Naturally only about half of this would become oxidized over the span of about ten years. The sequestration rate is then catalyzed to a rate of around 50 times the natural rate, thus eliminating the mangroves ability to absorb atmospheric carbon.

References

- Marine-ecosystem: biologydictionary.net, Retrieved 29 March, 2019

- Physical-and-chemical-properties-of-seawater, marine-ecosystem, science: britannica.com, Retrieved 30 April, 2019

- Ecosystem-shoreline: sciencing.com, Retrieved 29 June, 2019

- Rocky-shore, marine, environment: basicbiology.net, Retrieved 14 July, 2019

- Corals-and-coral-reefs, invertebrates, ocean-life: ocean.si.edu, Retrieved 11 January, 2019

- Deep-sea, ecosystems: ocean.si.edu, Retrieved 18 April, 2019

Chapter 4

Ocean Phenomena

There are numerous phenomena which occur within the ocean ecosystem. Some of these are sea foam, tidal bores, red tide, upwelling, whirlpool, etc. These phenomena have been discussed in detail in the following chapter.

The oceans contain great mysteries within their depths. While many of these mysteries have been explained by scientists and analysts, there are still quite a few unexplained oceanic enigmas that intrigue us. A variety of mysterious ocean phenomena have been seen and experienced by sailors around the world.

These various unexplained mysteries of the oceanic domain thus become interesting subjects of discussion and debate. Some such oceanic enigmas – both solved and unsolved ones – that have been popular topics of verbal analysing can be itemised as follows:

Milky Sea Phenomenon

Milky Sea refers to the unique milky glow of the waters of the Indian Ocean. The ocean phenomenon occurs on account of bioluminescent bacterial action and in turn, causes the water to turn blue, which appears to the naked eye as being milky white in colour in the darkness. The Milky Sea phenomenon has been documented to be in existence for over four centuries.

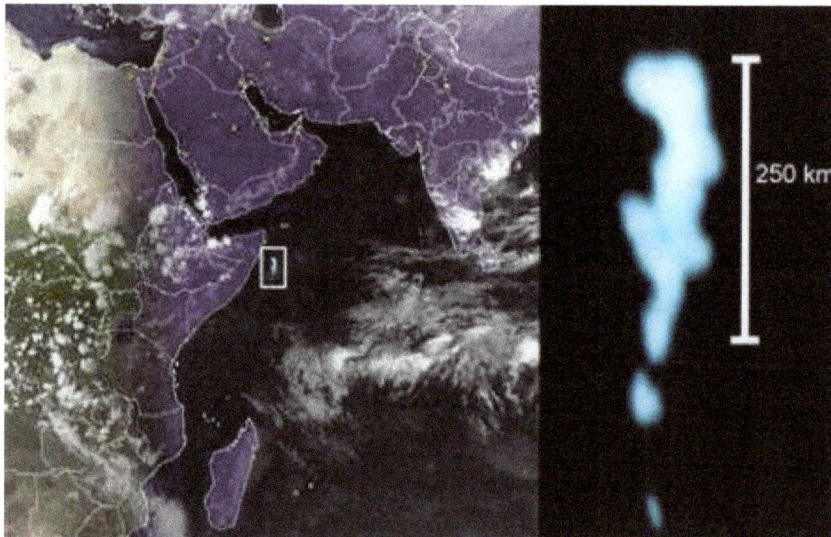

Bioluminescence

Bioluminescence is the light produced by marine creatures as a defence mechanism. Certain chemicals in the creatures' body when counteracted with atmospheric oxygen results in the emergence of bioluminescent light.

Convergence of Baltic and North Seas

This oceanic phenomenon has been a highly debated topic. The convergent point of the North and the Baltic Seas occurs in the province of Skagen in Denmark. However, because of the differing rates of densities of the seas' waters, the sea waters continue to remain separate in spite of their convergence. It is said that this ocean phenomenon finds a mention in the holy quran.

Steaming Black Sea

Called as the 'sea smoke', the steam arising out of the Black Sea is caused due to the humidity of the oceanic water counteracting with the coolness of the wind over the water's surface. Apart from explaining the ocean mystery behind the steam rising from the Black Sea, experts have also proved that the phenomenon is quite common to even smaller water bodies.

Green Flash

The ocean phenomenon of green flashes occurs during sunset and sunrise. Usually seen for merely a couple of seconds, such green flashes are the result of the natural prismatic effect of the atmosphere of the earth. During sunset and during sunrise, the light cast by the sun gets diverged into multiple colours, which is seen by the emitting of the green flash.

Baltic Sea Anomaly

The Baltic Sea Anomaly was accidentally discovered by a team of diving experts in the year 2011. The divers found a 60-metre thick circular entity nearly at a depth of 90 metres in the Baltic Sea.

A track seemed to lead towards the entity, which the divers measured to be around 300-metres. Though various scientists have offered innumerable suggestions about the entity's origins, the Baltic Sea anomaly still remains one of the unsolved intrigues and ocean mysteries of the world.'

Brinicle

Concentrated salt water escapes from within the frozen ice formed above the ocean's surface and seeps into the depths of the water. However, once the concentrated salt goes under the surface of the water, on account of natural processes it freezes and gets formed into brinicles. Brinicles occur in the frigid oceanic waters around the poles.

Underwater Crop Circle

Once regarded to be objects of high intrigue, the underwater crop circles have been explained to be a creative demonstration of pufferfishes' quests for finding their mates. These underwater circles have circumferences of over six feet and are often decorated with shells and other decorative items found at the bottom of the sea. The underwater crop circles were discovered under the waters of the Japanese island of Anami Oshima. Some consider these ocean mysteries as the work of aliens.

Red Tide

A red tide occurs when certain types of algae—plant-like organisms that live in the water—grow out of control. The name "red tide" comes from the fact that overgrowth of algae can cause the color of the water to turn red, as well as green or brown.

A red tide blooms off the coast of Texas.

Causes of Red Tide

Red tides are caused by algae, which are tiny, microscopic organisms that grow in the water. Almost all bodies of water have some algae, but in a red tide, there is a lot more algae in the water than usual. In fact, the water changes color in a red tide because the population of algae living in the water becomes so dense. Red tides have been around since long before humans. However, certain human activities are making them more frequent.

Chemicals from farming, factories, sewage treatment plants and other sources can become dissolved in water on the land. This water, called runoff, eventually flows into the ocean and can cause algae to grow faster, leading to red tides.

Nutrient-filled water, called runoff, can flow into lakes and oceans, contributing to algal blooms such as red tides. However, certain farming practices can reduce the amount of runoff that flows into streams and rivers, thus helping to prevent red tides.

Why are Red Tides Dangerous

Red tides are sometimes also called harmful algal blooms. Some of the algae that causes a red tide produce powerful toxins, which are harmful chemicals that can kill fish, shellfish, mammals and birds.

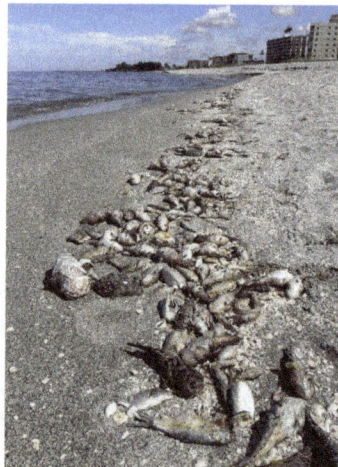

If people eat fish or shellfish that have been in the water with toxic algae, they will also ingest the toxins, which can make them sick. Many regions restrict fishing during a red tide for this reason. Nearby restaurants take local fish and shellfish off the menu, too.

Other types of harmful algal blooms are caused by nontoxic algae species that still make trouble. For example, when giant masses of algae bloom, they eventually die and start to decompose. As they decay, the oxygen levels in the water begin to decrease. The water can become so low in oxygen that animals in the water either swim away to healthier waters or die off. During a red tide, beaches are sometimes covered in dead fish and other animals that either ingested toxins or couldn't get enough oxygen.

Sea Foam

Sea foam, also referred to as ocean foam, beach foam, or spume is a type of foam created by the agitation of seawater, particularly when it contains higher concentrations of dissolved organic matter (proteins, fats, dead algae). These substances can act as surfactants or foaming agents. As the seawater is churned by breaking waves in the surf zone adjacent to the shore, the presence of these surfactants under turbulent conditions traps air, forming persistent bubbles which stick to each other through surface tension. Due to its low density and persistence, under some circumstances foam can be blown by strong on-shore winds from nearshore waters onto the beach itself and adjacent land.

Sea Foam at Fletcher Cove, Solana Beach.

Algal blooms are one common source of thick sea foams. When large blooms of algae decay offshore, great amounts of decaying algal matter often wash ashore. Foam forms as this organic matter is churned up by the surf.

Most sea foam is not harmful to humans and is often an indication of a productive ocean ecosystem. However, when large harmful algal blooms decay near shore, there is potential for impacts to human health and the environment. Along Gulf coast beaches during blooms of Karenia brevis, for example, popping sea foam bubbles are one way that algal toxins become airborne. The resulting aerosol can irritate the eyes of beach goers and poses a health risk for those with asthma or other respiratory conditions. Scientists studying the cause of a seabird die-offs off California in 2007 and in the Pacific Northwest in 2009 also found a soap-like foam from a decaying Akashiwo sanguinea algae bloom had removed the waterproofing on feathers, making it harder for birds to fly. This led to the onset of fatal hypothermia in many birds.

Sea foam near San Francisco.

Upwelling

Upwelling is a phenomenon that occurs in the ocean when the strong wind drives cooler, denser water from a lower surface in the ocean to the upper surface. In the process, the warmer water which is on the surface is moved to the bottom and is replaced by the cooler waters. The cooler, dense water that is moved during the upwelling process is usually rich in nutrients, unlike the water on the surface whose nutrients are usually exhausted by marine life. The intensity of the process depends on how strong the wind is. In some regions, the upwelling process is seasonal. In these regions, the marine productivity is also seasonal.

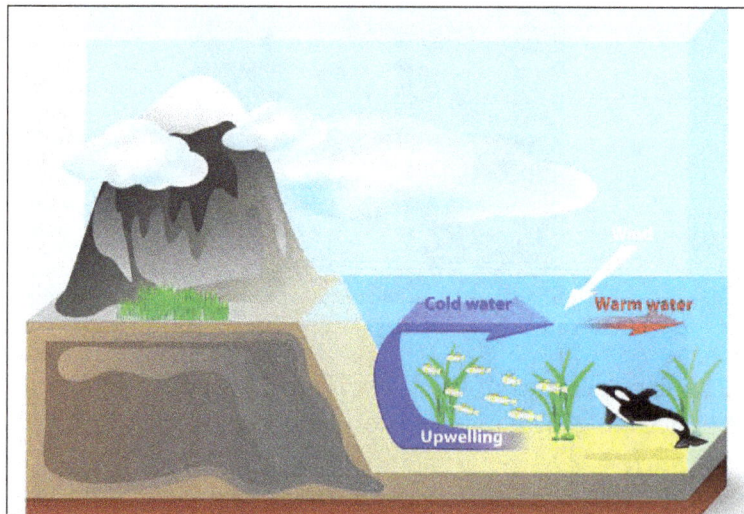

Upwelling drives cooler, denser water from the
lower surface of the ocean to the upper surface.

Mechanisms Behind Upwelling

The three mechanisms behind the upwelling process include the winds, the Ekman transport, and the Coriolis effects. The three mechanisms are important in the occurrences of different forms of upwelling. Generally, the wind blows across the water surface, leading to the water mixing with the wind, and eventually leading to the upwelling process. The wind results in the transportation of water at a rate of 90 degrees away from the wind's direction, a phenomenon brought about by Coriolis effects and Ekman transport. The Ekman transport is responsible for the approximately 45 degrees movement of water layer on the surface from the direction of the wind. The movement of water causes friction between the topmost layer and the layer below. This friction causes the subsequent layers to move in the similar direction as the topmost layer, resulting in a spiral-like movement of water. The Coriolis forces are responsible for the direction to which the water moves. If the upwelling is occurring in the northern hemisphere, the Coriolis force moves the water to the right-hand side of the wind while in the southern hemisphere the water is moved to the left-hand side of the direction of the wind.

Types

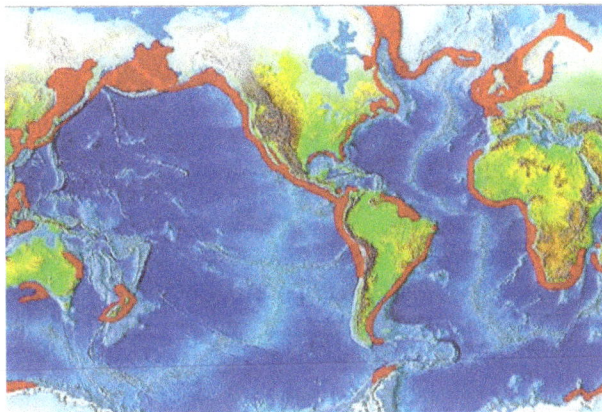

Areas of upwelling in red.

The major upwellings in the ocean are associated with the divergence of currents that bring deeper, colder, nutrient rich waters to the surface. There are at least five types of upwelling: coastal upwelling, large-scale wind-driven upwelling in the ocean interior, upwelling associated with eddies, topographically-associated upwelling, and broad-diffusive upwelling in the ocean interior.

Coastal

Coastal upwelling is the best known type of upwelling, and the most closely related to human activities as it supports some of the most productive fisheries in the world. Wind-driven currents are diverted to the right of the winds in the Northern Hemisphere and to the left in the Southern Hemisphere due to the Coriolis effect. The result is a net movement of surface water at right angles to the direction of the wind, known as the Ekman transport. When Ekman transport is occurring away from the coast, surface waters moving away are replaced by deeper, colder, and denser water. Normally, this upwelling process occurs at a rate of about 5–10 meters per day, but the rate and proximity of upwelling to the coast can be changed due to the strength and distance of the wind.

Deep waters are rich in nutrients, including nitrate, phosphate and silicic acid, themselves the result of decomposition of sinking organic matter (dead/detrital plankton) from surface waters. When brought to the surface, these nutrients are utilized by phytoplankton, along with dissolved CO_2 (carbon dioxide) and light energy from the sun, to produce organic compounds, through the process of photosynthesis. Upwelling regions therefore result in very high levels of primary production (the amount of carbon fixed by phytoplankton) in comparison to other areas of the ocean. They account for about 50% of global marine productivity. High primary production propagates up the food chain because phytoplankton are at the base of the oceanic food chain.

The food chain follows the course of:

Phytoplankton → Zooplankton → Predatory zooplankton → Filter feeders → Predatory fish → Marine birds, marine mammals.

Coastal upwelling exists year-round in some regions, known as major coastal upwelling systems, and only in certain months of the year in other regions, known as seasonal coastal upwelling systems. Many of these upwelling systems are associated with a relatively high carbon productivity and hence are classified as Large Marine Ecosystems.

Worldwide, there are five major coastal currents associated with upwelling areas: the Canary Current (off Northwest Africa), the Benguela Current (off southern Africa), the California Current (off California and Oregon), the Humboldt Current (off Peru and Chile), and the Somali Current (off Somalia and Oman). All of these currents support major fisheries. The four major eastern boundary currents in which coastal upwelling primarily occurs are the Canary Current, Benguela Current, California Current, and Humboldt Current. The Benguela Current is the eastern boundary of the South Atlantic subtropical gyre and can be divided into a northern and southern sub-system with upwelling occurring in both areas. The subsystems are divided by an area of permanent upwelling off of Luderitz, which is the strongest upwelling zone in the world. The California Current System (CCS) is an eastern boundary current of the North Pacific that is also characterized by a north and south split. The split in this system occurs at Point Conception, California due to weak upwelling in the South and strong upwelling in the north. The Canary Current is an eastern boundary current of the North Atlantic Gyre and is also separated due to the presence of the Canary Islands. Finally, the Humboldt Current or the Peru Current flows west along the coast of South America from Peru to Chile and extends up to 1,000 kilometers offshore. These four eastern boundary currents comprise the majority of coastal upwelling zones in the oceans.

Equatorial

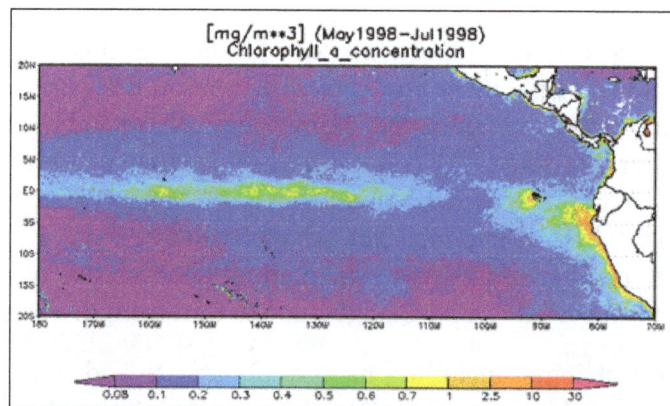

Effects of equatorial upwelling on surface chlorophyll concentrations in the Pacific ocean.

Upwelling at the equator is associated with the Intertropical Convergence Zone (ITCZ) which actually moves, and consequently, is often located just north or south of the equator. Easterly (westward) trade winds blow from the Northeast and Southeast and converge along the equator blowing West to form the ITCZ. Although there are no Coriolis forces present along the equator, upwelling still occurs just north and south of the equator. This results in a divergence, with denser, nutrient-rich water being upwelled from below, and results in the remarkable fact that the equatorial region in the Pacific can be detected from space as a broad line of high phytoplankton concentration.

Southern Ocean

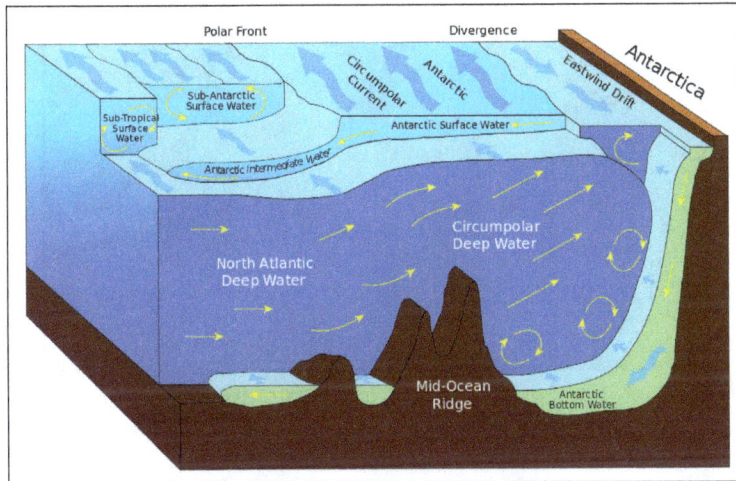

Upwelling in the Southern Ocean.

Large-scale upwelling is also found in the Southern Ocean. Here, strong westerly (eastward) winds blow around Antarctica, driving a significant flow of water northwards. This is actually a type of coastal upwelling. Since there are no continents in a band of open latitudes between South America and the tip of the Antarctic Peninsula, some of this water is drawn up from great depths. In many numerical models and observational syntheses, the Southern Ocean upwelling represents the primary means by which deep dense water is brought to the surface. In some regions of Antarctica, wind-driven upwelling near the coast pulls relatively warm Circumpolar deep water onto the continental shelf, where it can enhance ice shelf melt and influence ice sheet stability. Shallower, wind-driven upwelling is also found in off the west coasts of North and South America, northwest and southwest Africa, and southwest and south Australia, all associated with oceanic subtropical high pressure circulations.

Some models of the ocean circulation suggest that broad-scale upwelling occurs in the tropics, as pressure driven flows converge water toward the low latitudes where it is diffusively warmed from above. The required diffusion coefficients, however, appear to be larger than are observed in the real ocean. Nonetheless, some diffusive upwelling does probably occur.

Other Sources

- Local and intermittent upwellings may occur when offshore islands, ridges, or seamounts cause a deflection of deep currents, providing a nutrient rich area in otherwise low productivity ocean areas. Examples include upwellings around the Galapagos Islands and the Seychelles Islands, which have major pelagic fisheries.

- Upwelling can also occur when a tropical cyclone transits an area, usually when moving at speeds of less than 5 mph (8 km/h). The churning of a cyclone eventually draws up cooler water from lower layers of the ocean. This causes the cyclone to weaken.

- Artificial upwelling is produced by devices that use ocean wave energy or ocean thermal energy conversion to pump water to the surface. Ocean wind turbines are also known to produce upwellings. Ocean wave devices have been shown to produce plankton blooms.

Variations

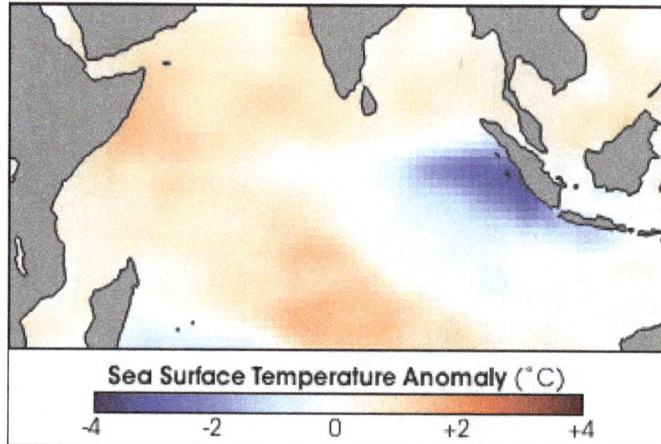

Unusually strong winds from the east push warm (red) surface water towards
Africa, allowing cold (blue) water to upwell along the Sumatran coast.

Upwelling intensity depends on wind strength and seasonal variability, as well as the vertical structure of the water, variations in the bottom bathymetry, and instabilities in the currents.

In some areas, upwelling is a seasonal event leading to periodic bursts of productivity similar to spring blooms in coastal waters. Wind-induced upwelling is generated by temperature differences between the warm, light air above the land and the cooler denser air over the sea. In temperate latitudes, the temperature contrast is greatly seasonably variable, creating periods of strong upwelling in the spring and summer, to weak or no upwelling in the winter. For example, off the coast of Oregon, there are four or five strong upwelling events separated by periods of little to no upwelling during the six-month season of upwelling. In contrast, tropical latitudes have a more constant temperature contrast, creating constant upwelling throughout the year. The Peruvian upwelling, for instance, occurs throughout most of the year, resulting in one of the world's largest marine fisheries for sardines and anchovies.

In anomalous years when the trade winds weaken or reverse, the water that is upwelled is much warmer and low in nutrients, resulting in a sharp reduction in the biomass and phytoplankton productivity. This event is known as the El Nino-Southern Oscillation (ENSO) event. The Peruvian upwelling system is particularly vulnerable to ENSO events, and can cause extreme interannual variability in productivity.

Changes in bathymetry can affect the strength of an upwelling. For example, a submarine ridge that extends out from the coast will produce more favorable upwelling conditions than neighboring regions. Upwelling typically begins at such ridges and remains strongest at the ridge even after developing in other locations.

High Productivity

Since upwelling regions are important sources of marine productivity, and they attract hundreds of species throughout the trophic levels, the diversity of these systems has been a focal point for marine research. While studying the trophic levels and patterns typical of upwelling regions, researchers have discovered that upwelling systems exhibit a wasp-waist richness pattern. In this type of pattern, the high and low trophic levels are well-represented by high species diversity. However, the intermediate trophic level is only represented by one or two species. This trophic layer, which consists of small, pelagic fish usually makes up about only three to four percent of the species diversity of all fish species present. The lower trophic layers are very well-represented with about 500 species of copepods, 2500 species of gastropods, and 2500 species of crustaceans on average. At the apex and near-apex trophic levels, there are usually about 100 species of marine mammals and about 50 species of marine birds. The vital intermediate trophic species however are small pelagic fish that usually feed on phytoplankton. In most upwelling systems, these species are either anchovies or sardines, and usually only one is present, although two or three species may be present occasionally. These fish are an important food source for predators, such as large pelagic fish, marine mammals, and marine birds. Although they are not at the base of the trophic pyramid, they are the vital species that connect the entire marine ecosystem and keep the productivity of upwelling zones so high.

Threats to Upwelling Ecosystems

A major threat to both this crucial intermediate trophic level and the entire upwelling trophic ecosystem is the problem of commercial fishing. Since upwelling regions are the most productive and species rich areas in the world, they attract a high number of commercial fishers and fisheries. On one hand, this is another benefit of the upwelling process as it serves as a viable source of food and income for so many people and nations besides marine animals. However, just as in any ecosystem, the consequences of over-fishing from a population could be detrimental to that population and the ecosystem as a whole. In upwelling ecosystems, every species present plays a vital role in the functioning of that ecosystem. If one species is significantly depleted, that will have an effect throughout the rest of the trophic levels. For example, if a popular prey species is targeted by fisheries, fishermen may collect hundreds of thousands of individuals of this species just by casting their nets into the upwelling waters. As these fish are depleted, the food source for those who preyed on these fish is depleted. Therefore, the predators of the targeted fish will begin to die off, and there will not be as many of them to feed the predators above them. This system continues throughout the entire food chain, resulting in a possible collapse of the ecosystem. It is possible that the ecosystem may be restored over time, but not all species can recover from events such as these. Even if the species can adapt, there may be a delay in the reconstruction of this upwelling community.

The possibility of such an ecosystem collapse is the very danger of fisheries in upwelling regions. Fisheries may target a variety of different species, and therefore they are a direct threat to many species in the ecosystem, however they pose the highest threat to the intermediate pelagic fish. Since these fish form the crux of the entire trophic process of upwelling ecosystems, they are highly represented throughout the ecosystem (even if there is only one species present). Unfortunately, these fish tend to be the most popular targets of fisheries as about 64 percent of their entire catch consists of pelagic fish. Among those, the six main species that usually form the intermediate trophic layer represent over half of the catch.

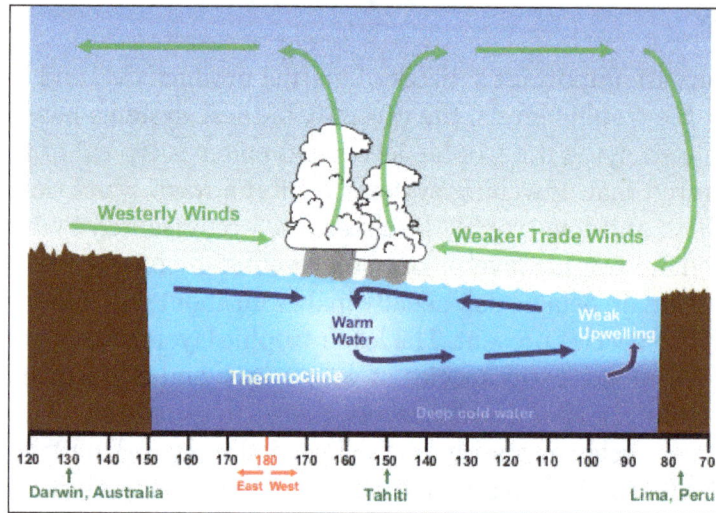

During an El Niño, wind indirectly drives warm water to the
South American coast, reducing the effects of cold upwelling.

Besides directly causing the collapse of the ecosystem due to their absence, this can create problems in the ecosystem through a variety of other methods as well. The animals higher in the trophic levels may not completely starve to death and die off, but the decreased food supply could still hurt the populations. If animals do not get enough food, it will decrease their reproductive viability meaning that they will not breed as often or as successfully as usual. This can lead to a decreasing population, especially in species that do not breed often under normal circumstances or become reproductively mature late in life. Another problem is that the decrease in the population of a species due to fisheries can lead to a decrease in genetic diversity, resulting in a decrease in bio-diversity of a species. If the species diversity is decreased significantly, this could cause problems for the species in an environment that is so variable and quick-changing; they may not be able to adapt, which could result in a collapse of the population or ecosystem.

Another threat to the productivity and ecosystems of upwelling regions is El Niño-Southern Oscillation (ENSO) system, or more specifically El Niño events. During the normal period and La Niña events, the easterly trade winds are still strong, which continues to drive the process of upwelling. However, during El Niño events, trade winds are weaker, causing decreased upwelling in the equatorial regions as the divergence of water north and south of the equator is not as strong or as prevalent. The coastal upwelling zones diminish as well since they are wind driven systems, and the wind is no longer a very strong driving force in these areas. As a result, global upwelling drastically decreases, causing a decrease in productivity as the waters are no longer receiving nutrient-rich water. Without these nutrients, the rest of the trophic pyramid cannot be sustained, and the rich upwelling ecosystem will collapse.

Tsunami

A tsunami is a series of waves created when a body of water, such as an ocean, is rapidly displaced. Earthquakes, mass movements above or below water, volcanic eruptions and other underwater explosions, landslides, large meteorite impacts, and nuclear weapons testing at sea all have the potential to generate a tsunami. A tsunami can have a range of effects, from unnoticeable to devastating.

A tsunami has a much smaller amplitude (wave height) offshore, and a very long wavelength (often hundreds of kilometers long). Consequently, they generally pass unnoticed at sea, forming only a passing "hump" in the ocean.

The tsunami that struck Malé in the Maldives.

Tsunami have been historically referred to as tidal waves because, as they approach land, they take on the characteristics of a violent, onrushing tide, rather than the sort of cresting waves formed by wind action on the ocean. Given that they are not actually related to tides, the term is considered misleading and its usage is discouraged by oceanographers.

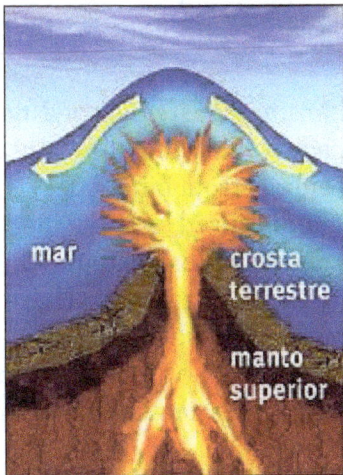

Volcanic eruptions inject tons of wash in the oceanic soil, generating devastating waves.

Submarine earthquakes dislocate the oceanic crust, pushing water upwards.

Causes

A tsunami can be generated when the plate boundaries abruptly deform and vertically displace the overlying water. Such large vertical movements of the Earth's crust can occur at plate boundaries. Subduction earthquakes are particularly effective in generating tsunami. Also, one tsunami in the 1940s in Hilo, Hawaii, was actually caused by an earthquake on one of the Aleutian Islands in Alaska. That earthquake was 7.8 on the Richter Scale.

Tsunami are formed as the displaced water mass moves under the influence of gravity and radiates across the ocean like ripples on a pond.

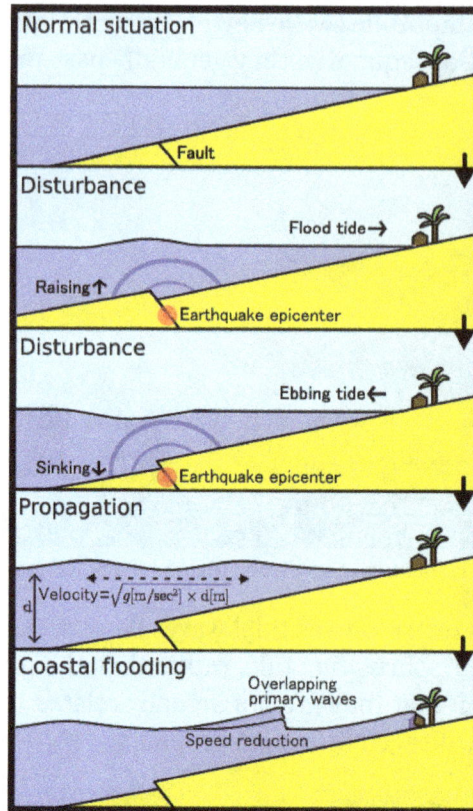

Generation of a tsunami.

In the 1950s, it was discovered that larger tsunami than previously believed possible could be caused by landslides, explosive volcanic action, and impact events when they contact water. These phenomena rapidly displace large volumes of water, as energy from falling debris or expansion is transferred to the water into which the debris falls. Tsunami caused by these mechanisms, unlike the ocean-wide tsunami caused by some earthquakes, generally dissipate quickly and rarely affect coastlines distant from the source due to the small area of sea affected. These events can give rise to much larger local shock waves (solitons), such as the landslide at the head of Lituya Bay which produced a water wave estimated at 50 – 150 m and reached 524 m up local mountains. However, an extremely large landslide could generate a "megatsunami" that might have ocean-wide impacts.

The geological record tells us that there have been massive tsunami in Earth's past.

Signs of an Approaching Tsunami

There is often no advance warning of an approaching tsunami. However, since earthquakes are often a cause of tsunami, an earthquake felt near a body of water may be considered an indication that a tsunami will shortly follow.

When the first part of a tsunami to reach land is a trough rather than a crest of the wave, the water along the shoreline may recede dramatically, exposing areas that are normally always submerged. This can serve as an advance warning of the approaching crest of the tsunami, although the warning arrives only a very short time before the crest, which typically arrives seconds to minutes later.

In the 2004 tsunami that occurred in the Indian Ocean, the sea receding was not reported on the African coast or any other western coasts it hit, when the tsunami approached from the east.

Tsunami occur most frequently in the Pacific Ocean, but are a global phenomenon; they are possible wherever large bodies of water are found, including inland lakes, where they can be caused by landslides. Very small tsunami, non-destructive and undetectable without specialized equipment, occur frequently as a result of minor earthquakes and other events.

Warnings and Prevention

A tsunami can also be known to come when the water leaves an ocean or large body of water, and then the water in it causes a large series of waves to approach land.

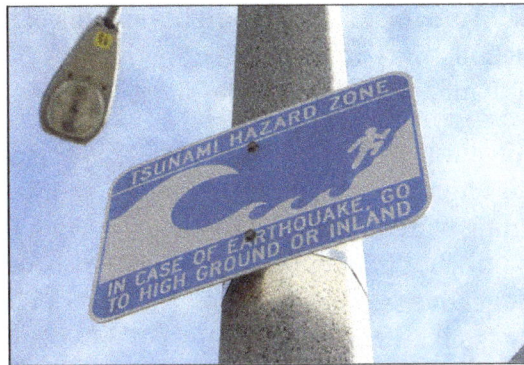

Tsunami hazard sign at The Wedge in Balboa Peninsula.

Tsunami cannot be prevented or precisely predicted, but there are some warning signs of an impending tsunami, and there are many systems being developed and in use to reduce the damage from tsunami.

In instances where the leading edge of the tsunami wave is its trough, the sea will recede from the coast half of the wave's period before the wave's arrival. If the slope is shallow, this recession can exceed many hundreds of meters. People unaware of the danger may remain at the shore due to curiosity, or for collecting shellfish from the exposed seabed.

Tsunami wall at Tsu, Japan.

Regions with a high risk of tsunami may use tsunami warning systems to detect tsunami and warn the general population before the wave reaches land. In some communities on the west coast of the United States, which is prone to Pacific Ocean tsunami, warning signs advise people where to run in the event of an incoming tsunami. Computer models can roughly predict tsunami arrival and impact based on information about the event that triggered it and the shape of the seafloor (bathymetry) and coastal land (topography).

One of the early warnings comes from nearby animals. Many animals sense danger and flee to higher ground before the water arrives. The Lisbon quake is the first documented case of such a phenomenon in Europe. The phenomenon was also noted in Sri Lanka in the 2004 Indian Ocean earthquake. Some scientists speculate that animals may have an ability to sense subsonic Rayleigh waves from an earthquake minutes or hours before a tsunami strikes shore). More likely, though, is that the certain large animals (e.g., elephants) heard the sounds of the tsunami as it approached the coast. The elephants' reactions were to go in the direction opposite of the noise, and thus go inland. Humans, on the other hand, head down to the shore to investigate.

While it is not possible to prevent tsunami, in some particularly tsunami-prone countries some measures have been taken to reduce the damage caused on shore. Japan has implemented an extensive programme of building tsunami walls of up to 4.5 m (13.5 ft) high in front of populated coastal areas. Other localities have built floodgates and channels to redirect the water from incoming tsunami. However, their effectiveness has been questioned, as tsunami are often higher than the barriers. For instance, the tsunami which struck the island of Hokkaidō on July 12, 1993 created waves as much as 30 m (100 ft) tall - as high as a ten-story building. The port town of Aonae was completely surrounded by a tsunami wall, but the waves washed right over the wall and destroyed all the wood-framed structures in the area. The wall may have succeeded in slowing down and moderating the height of the tsunami, but it did not prevent major destruction and loss of life.

Whirlpool

A whirlpool is a swirling water that is formed when two opposing currents meet. A whirlpool can either be very powerful and strong or not very powerful and small. Powerful ones are often referred to as maelstroms and are mainly common in seas and oceans. Smaller whirlpools are common at the base of waterfalls and can also be observed in man-made structures such as dams and weirs. In oceans, they are mainly caused by tides and are capable of submerging large ships. Whirlpool that has a downdraft is referred to as Vortex in proper terms. Below are some of the largest and notable whirlpools in the world.

Corryvreckan

The Gulf of Corryvreckan is a strait located between Jura and Scarba islands, Scotland. The underwater topography of the area and the strong Atlantic current conspire to create a strong tidal race. As the tides enter the strait it speeds up and meet a variety of seabed features such as deep holes and rising pinnacles. The features come together to form the Corryvreckan whirlpool, the third-largest in the world. The whirlpool is located on the northern part of the gulf and surrounds the pyramid-shaped pinnacle. The Corryvreckan Whirlpool is believed to be unnavigable by many although the Royal Navy has not classified it as so. However, the nearby Little Corryvreckan has been officially classified as unnavigable because of its violence. The divers who explore the area consider it as one of the most dangerous dives in Britain.

Naruto Whirlpools

Naruto Strait located between Naruto and Awaji Island in Tokushima and Hyogo, Japan respectively is known for a tidal whirlpool which has been named Naruto Whirlpool. The strait is 0.81 miles wide and connects the Inland Sea to the Pacific Ocean. The tide causes a large amount of water to move in and out of the Inland Sea twice in a day. The tide, with a range of up to 5.6 feet, causes a difference in water level of approximately 4 feet between the Inland Sea and the Pacific Ocean. Because the Naruto Strait is narrow, water rushes into it through the Naruto Channel at about 8 mph four times in a day and at 12 mph during spring tide, creating vortices of up to 66 feet in diameter. The current at the Naruto Strait is Japan's fastest and 4th fastest in the world. Naruto Bridge offers the perfect spot for observing the whirlpools.

Old Sow

Old Sow is one of the largest whirlpools in the Western Hemisphere. It is situated off the shores of Deer Island in New Brunswick. The phrase "Old Sow" is obtained from the pig-like noise made by the whirlpool. It is formed by an extreme tidal range where there is an exchange of water between the Bay of Fundy and Passamaquoddy Bay and is accelerated by the unusual seafloor topography in the area. The whirlpool forms in a region of about 250 feet in diameter. The tremendous water turbulence takes place locally on the larger Old Sow region, frequently forming a huge area of several fascinating turbulence. The activities of Old Sow whirlpool is affected by the numerous currents and counter currents including tidal surges, storms, and strong winds. The whirlpool can be best viewed at Deer Point on Deer Island at the south end of the island or from Moose Island. Several other small whirlpools can also be seen in the area.

Skookumchuck Narrows

Skookumchuck Narrows forms the entrance of Sechelt Inlet on Sunshine Coast in British Columbia, Canada. The tidal flow of the strait and those of the Salmon Inlet and Narrows Inlet pass through the Sechelt Rapid before broadening into Sechelt Inlet. Whitecaps and whirlpools are often formed at the rapids during peak flows. Every day, large amounts of seawater are forced through the Skookumchuck Narrows by the tides. The water level difference on either side of the rapid may exceed 6.6 feet in height while the current speed may exceed 30 km/h. The tidal rapids are sometimes considered the fastest in the world. The tidal pattern causes the water to move almost all times in the narrows area.

Mokstraumen

Mokstraumen is one of the strongest whirlpools in the world. It is also a system tidal eddies that form on the Norwegian Sea, between the Lofoten Point of Moskenes and Vaeroy on Mosken Island. The whirlpool is formed when the strong tidal current flow between the islands and the Atlantic Ocean and the deep Vestfjorden. The largest whirlpool has a diameter of 130 to 160 feet and induces a surface water ripple of up to 3 feet. Mokstraumen result from several factors such as tides, strong winds, the position of the Lofotodden, and the topography of the underwater. Although most whirlpools occur in confined straits and rivers, Mokstraumen is an exception. It occurs in open seas. Tides on Lofoten rise twice a day and are the major contributors to the whirlpool. They combine with the Norwegian Sea current and storm-induced flow, resulting in a significant stream. The flow occurs at a depth of about 1,600 feet and meets a ridge of about 66 feet deep at Mosken and Vaeroy islands, leading to an upward movement and eddies around the region.

Saltstraumen

Saltstraumen is a tiny strait with a strong tidal current. The strait is situated in the municipality of Bodo in Nordland County, Norway, about 6.2 miles southeast of Bodo Town. It is about 1.9 miles long and 490 feet wide. About 400 million cubic meters of water forces its way through Saltstraumen every six hours. Whirlpools of up to 33 feet in diameter and 16 feet in depth are often formed in the area when the currents are strongest. The current is formed when the tides try to fill

Skjerstand Fjord. The height difference between the sea level and fjord inside can be up to 3 feet. Saltstraumen has abundant fish including cod, wolfish, and saithe. The most common species is the coalfish with some as large as 50 pounds.

Maelstrom

A maelstrom is a powerful whirlpool. With a name like maelstrom, it's hard to ignore the way that this phenomenon kind of warns us about it's dangerous ways. Originally introduced to the world by author Edgar Allan Poe, the word maelstrom actually means "crushing current" which is very accurate. It is essentially a powerful and vast whirlpool that has a downdraft that would quickly suck up whatever was in the immediate vicinity. The weather is something that is a huge determination of the force and strength of a maelstrom. There are urban legends that insist that a maelstrom will immediately lead you to the bottom of the ocean, but those are dismantled by scientists. Who would even want to go to the bottom of the middle of the ocean anyway? Seems like a scary place.

Tidal Bores

A tidal bore is a phenomenon in which the leading edge of the incoming tide forms a wave (or waves) of water that travels up a river or narrow bay against the direction of the river or bay's current.

Bores occur in relatively few locations worldwide, usually in areas with a large tidal range (typically more than 6 meters (20 ft) between high and low water) and where incoming tides are funneled into a shallow, narrowing river or lake via a broad bay. The funnel-like shape not only increases the tidal range, but it can also decrease the duration of the flood (incoming) tide, down to a point where the flood appears as a sudden increase in the water level.

A tidal bore may take on various forms, ranging from a single breaking wavefront with a roller — somewhat like a hydraulic jump — to "undular bores", comprising a smooth wavefront followed by a train of secondary waves (whelps). Large bores can be particularly unsafe for shipping but also present opportunities for river surfing.

Two key features of a tidal bore are the intense turbulence and mixing generated during the bore propagation, as well as its rumbling noise. A tidal bore creates a powerful roar that combines the sounds caused by the turbulence in the bore front and whelps, entrained air bubbles in the bore roller, sediment erosion beneath the bore front and of the banks, scouring of shoals and bars, and impacts on obstacles.

Regional Distribution Tidal Bore

Tidal bore can be found almost every part of the world. It can occur in rivers and lakes. The lakes, rivers and bays that have been known to exhibit bores are given below:

- Ganges, Brahmaputra, Indus River, Sittaung River, Burma, Qiantang River, Batang Lupar or Lupar River, Batang Sadong or Sadong River, Bono, Kampar River.

- Styx River and Daly River.

- River Shannon, River Trent, River Dee, River Mersey, River Severn, River Parrett, River Welland, River Kent, River Great Ouse, River Ouse, River Eden, River Esk, River Nith, River Lune, Lancashire, River Ribble, River Yealm, Devon, River Leven, River Ribble, Durme, river training, Bay of Mont-Saint-Michel, Arguenon, Baie de la Frênaye, Vire, Sienne, Vilaine (locally named le mascarin), Dordogne and Garonne.

- Fly River and Turama River.

- Petitcodiac River, Turnagain arm of Cook Inlet (Alaska), Colorado River, Savannah River, Mississippi Gulf Coast, Bay of Fundy (between Nova Scotia and New Brunswick), Petitcodiac River, Shubenacadie River, River Hebert and Maccan River, the St. Croix and Kennetcook rivers, the Salmon River, Colorado River.

- Amazon River, Orinoco River, Mearim River and Araguari River.

Effects of Tidal Bores

As we know that tidal bore is a combination of forces and activities including turbulence in the whelps, as well as impacts on other obstacles on the riverfront. Therefore, it is considered as dangerous because these waves are turbulence and rumble is driven, which interfere the navigation routes and shipping activities. It also provides ample space for rich breeding and spawning ground for fishes and to support the shrimp species through inducing aeration.

References

- Top-10-amazing-ocean-mysteries-and-phenomena, environment: marineinsight.com, Retrieved 27 May, 2019

- Sea-foam: beachapedia.org, Retrieved 21 July, 2019

- What-is-upwelling: worldatlas.com, Retrieved 13 July, 2019

- Tsunami, entry: newworldencyclopedia.org, Retrieved 24 March , 2019

- The-world-s-largest-whirlpools: worldatlas.com, Retrieved 1 June, 2019

- Top-10-incredible-ocean-phenomena: toptenz.net, Retrieved 12 April, 2019

- Tidal-bore: beachapedia.org, Retrieved 23 June, 2019

Chapter 5

Diverse Aspects of Ocean

Some of the diverse aspects of the ocean studied within oceanography are ocean current, tides and waves. It also studies ocean basins and sediments. The topics elaborated in this chapter will help in gaining a better perspective about these diverse aspects of the ocean.

Ocean Current

Ocean current is stream made up of horizontal and vertical components of the circulation system of ocean waters that is produced by gravity, wind friction, and water density variation in different parts of the ocean. Ocean currents are similar to winds in the atmosphere in that they transfer significant amounts of heat from Earth's equatorial areas to the poles and thus play important roles in determining the climates of coastal regions. In addition, ocean currents and atmospheric circulation influence one another.

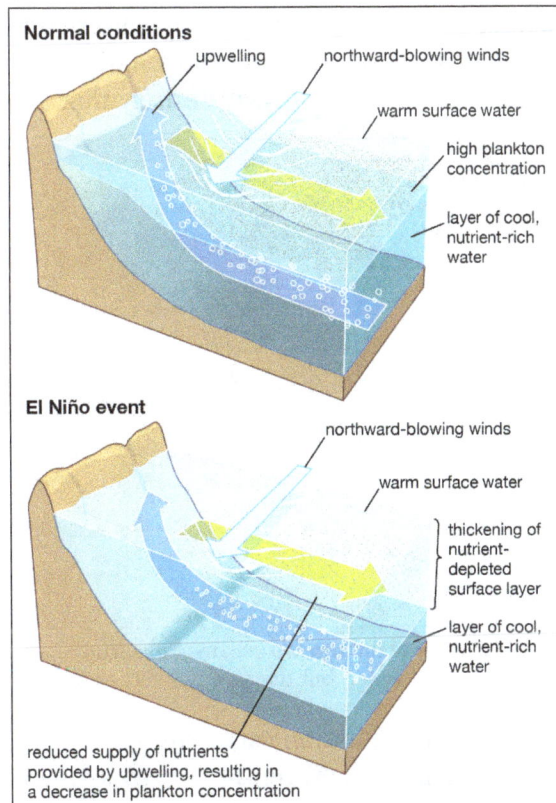

The upwelling process in the ocean along the coast of Peru. A thermocline and a nutricline separate the warm, nutrient-deficient upper layer from the cool, enriched layer below. Under normal conditions (top), these interfaces are shallow enough that coastal winds can induce upwelling of the lower-layer nutrients to the surface, where they support an abundant ecosystem. During an El Niño event (bottom), the upper layer thickens so that the upwelled water contains fewer nutrients, thus contributing to a collapse of marine productivity.

The general circulation of the oceans defines the average movement of seawater, which, like the atmosphere, follows a specific pattern. Superimposed on this pattern are oscillations of tides and waves, which are not considered part of the general circulation. There also are meanders and eddies that represent temporal variations of the general circulation. The ocean circulation pattern exchanges water of varying characteristics, such as temperature and salinity, within the interconnected network of oceans and is an important part of the heat and freshwater fluxes of the global climate. Horizontal movements are called currents, which range in magnitude from a few centimetres per second to as much as 4 metres (about 13 feet) per second. A characteristic surface speed is about 5 to 50 cm (about 2 to 20 inches) per second. Currents generally diminish in intensity with increasing depth. Vertical movements, often referred to as upwelling and downwelling, exhibit much lower speeds, amounting to only a few metres per month. As seawater is nearly incompressible, vertical movements are associated with regions of convergence and divergence in the horizontal flow patterns.

Distribution of Ocean Currents

Maps of the general circulation at the sea surface were originally constructed from a vast amount of data obtained from inspecting the residual drift of ships after course direction and speed are accounted for in a process called dead reckoning. This information is collected by satellite-tracked surface drifters at sea at present. The pattern is nearly entirely that of wind-driven circulation.

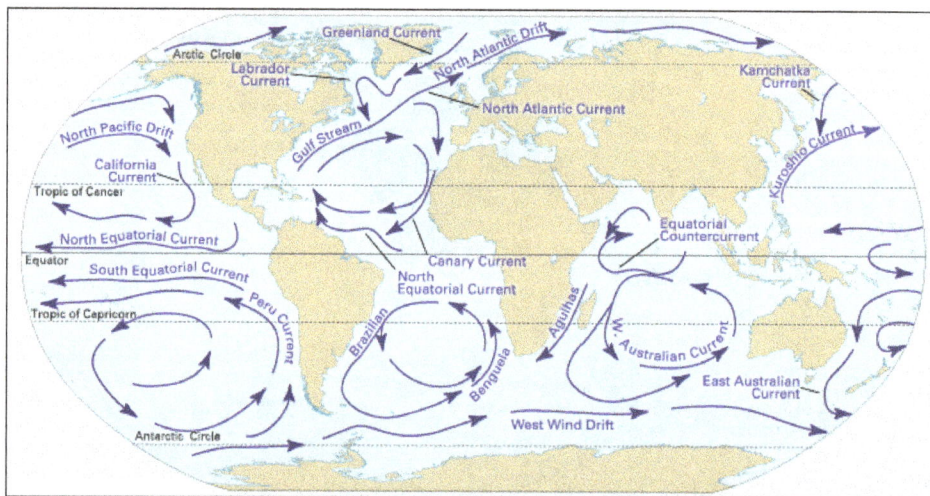

Major surface currents of the world's oceans: Subsurface currents also move vast amounts of water, but they are not known in such detail.

At the surface, aspects of wind-driven circulation cause the gyres (large anticyclonic current cells that spiral about a central point) to displace their centres westward, forming strong western boundary currents against the eastern coasts of the continents, such as the Gulf Stream–North Atlantic–Norway Current in the Atlantic Ocean and the Kuroshio–North Pacific Current in the Pacific Ocean. In the Southern Hemisphere the counterclockwise circulation of the gyres creates strong eastern boundary currents against the western coasts of continents, such as the Peru (Humboldt) Current off South America, the Benguela Current off western Africa, and the Western Australia Current. The Southern Hemisphere currents are also influenced by the powerful, eastward-flowing, circumpolar Antarctic Current. It is a very deep, cold, and relatively slow current, but it carries a vast mass of water, about twice the volume of the Gulf Stream. The Peru and Benguela currents draw water from this Antarctic current and, hence, are cold. The Northern Hemisphere lacks continuous open water bordering the Arctic and so has no corresponding powerful circumpolar current, but there are small cold currents flowing south through the Bering Strait to form the Oya and Anadyr currents off eastern Russia and the California Current off western North America; others flow south around Greenland to form the cold Labrador and East Greenland currents. The Kuroshio–North Pacific and Gulf Stream–North Atlantic–Norway currents move warmer water into the Arctic Ocean via the Bering, Cape, and West Spitsbergen currents.

In the tropics the great clockwise and counterclockwise gyres flow westward as the Pacific North and South Equatorial currents, Atlantic North and South Equatorial currents, and the Indian South Equatorial Current. Because of the alternating monsoon climate of the northern Indian Ocean, the current in the northern Indian Ocean and the Arabian Sea alternates. Between these massive currents are narrow eastward-flowing countercurrents.

Other smaller current systems found in certain enclosed seas or ocean areas are less affected by wind-driven circulation and more influenced by the direction of water inflow. Such currents are found in the Tasmanian Sea, where the southward-flowing East Australian Current generates counterclockwise circulation, in the northwestern Pacific, where the eastward-flowing Kuroshio–North Pacific current causes counterclockwise circulation in the Alaska Current and Aleutian Current (or Subarctic Current), in the Bay of Bengal, and in the Arabian Sea.

Deep-ocean circulation consists mainly of thermohaline circulation. The currents are inferred from the distribution of seawater properties, which trace the spreading of specific water masses. The distribution of density is also used to estimate the deep currents. Direct observations of subsurface currents are made by deploying current meters from bottom-anchored moorings and by setting out neutral buoyant instruments whose drift at depth is tracked acoustically.

Causes of Ocean Currents

The general circulation is governed by the equation of motion, one of the fundamental laws of mechanics developed by English physicist and mathematician Sir Isaac Newton that was applied to a continuous volume of water. This equation states that the product of mass and current acceleration equals the vector sum of all forces that act on the mass. Besides gravity, the most important forces that cause and affect ocean currents are horizontal pressure-gradient forces, Coriolis forces, and frictional forces. Temporal and inertial terms are generally of secondary importance to the general flow, though they become important for transient features such as meanders and eddy.

Pressure Gradients

The hydrostatic pressure, p, at any depth below the sea surface is given by the equation $p = g\rho z$, where g is the acceleration of gravity, ρ is the density of seawater, which increases with depth, and z is the depth below the sea surface. This is called the hydrostatic equation, which is a good approximation for the equation of motion for forces acting along the vertical. Horizontal differences in density (due to variations of temperature and salinity) measured along a specific depth cause the hydrostatic pressure to vary along a horizontal plane or geopotential surface, a surface perpendicular to the direction of the gravity acceleration. Horizontal gradients of pressure, though much smaller than vertical changes in pressure, give rise to ocean currents.

In a homogeneous ocean, which would have a constant potential density, horizontal pressure differences are possible only if the sea surface is tilted. In this case, surfaces of equal pressure, called isobaric surfaces, are tilted in the deeper layers by the same amount as the sea surface. This is referred to as the barotropic field of mass. The unchanged pressure gradient gives rise to a current speed independent of depth. The oceans of the world, however, are not homogeneous. Horizontal variations in temperature and salinity cause the horizontal pressure gradient to vary with depth. This is the baroclinic field of mass, which leads to currents that vary with depth. The horizontal pressure gradient in the ocean is a combination of these two mass fields.

The tilt, or topographic relief, of the isobaric surface marking sea surface (defined as p = 0) can be constructed from a three-dimensional density distribution using the hydrostatic equation. Since the absolute value of pressure is not measured at all depths in the ocean, the sea surface slope is presented relative to that of a deep isobaric surface; it is assumed that the deep isobaric surface is level. Since the wind-driven circulation attenuates with increasing depth, an associated decrease of isobaric tilt with increasing depth is expected. Representation of the sea surface relief relative to a deep reference surface is a good representation of the absolute shape of the sea surface. The total relief of the sea surface amounts to about 2 metres (about 6.5 feet), with "hills" in the subtropics and "valleys" in the polar regions. This pressure head drives the surface circulation.

Coriolis Effect

Earth's rotation about its axis causes moving particles to behave in a way that can only be understood by adding a rotational dependent force. To an observer in space, a moving body would continue to move in a straight line unless the motion were acted upon by some other force. To an Earth-bound observer, however, this motion cannot be along a straight line because the reference frame is the rotating Earth. This is similar to the effect that would be experienced by an observer standing on a large turntable if an object moved over the turntable in a straight line relative to the "outside" world. An apparent deflection of the path of the moving object would be seen. If the turntable rotated counterclockwise, the apparent deflection would be to the right of the direction of the moving object, relative to the observer fixed on the turntable.

This remarkable effect is evident in the behaviour of ocean currents. It is called the Coriolis force, named after Gustave-Gaspard Coriolis, a 19th-century French engineer and mathematician. For Earth, horizontal deflections due to the rotational induced Coriolis force act on particles moving in any horizontal direction. There also are apparent vertical forces, but these are of minor importance to ocean currents. Because Earth rotates from west to east about its axis, an observer in the Northern Hemisphere would notice a deflection of a moving body toward the right. In the Southern Hemisphere, this deflection would be toward the left. As a result, ocean currents move clockwise (anticyclonically) in the Northern Hemisphere and counterclockwise (cyclonically) in the Southern Hemisphere; Coriolis force deflects them about 45° from the wind direction, and at the Equator there would be no apparent horizontal deflection.

It can be shown that the Coriolis force always acts perpendicular to motion. Its horizontal component, C_f, is proportional to the sine of the geographic latitude (θ, given as a positive value for the Northern Hemisphere and a negative value for the Southern Hemisphere) and the speed, c, of the moving body. It is given by $C_f = c\,(2\omega \sin \theta)$, where $\omega = 7.29 \times 10^{-5}$ radian per second is the angular velocity of Earth's rotation.

Frictional Forces

Movement of water through the oceans is slowed by friction, with surrounding fluid moving at a different velocity. A faster-moving fluid layer tends to drag along a slower-moving layer, and a slower-moving layer will tend to reduce the speed of a faster-moving layer. This momentum transfer between the layers is referred to as frictional forces. The momentum transfer is a product of turbulence that moves kinetic energy to smaller scales until at the tens-of-microns scale (1 micron = 1/1,000 mm) it is dissipated as heat. The wind blowing over the sea surface transfers momentum to the water. This frictional force at the sea surface (i.e., the wind stress) produces the wind-driven circulation. Currents moving along the ocean floor and the sides of the ocean also are subject to the influence of boundary-layer friction. The motionless ocean floor removes momentum from the circulation of the ocean waters.

Geostrophic Currents

For most of the ocean volume away from the boundary layers, which have a characteristic thickness of 100 metres (about 330 feet), frictional forces are of minor importance, and the equation of motion for horizontal forces can be expressed as a simple balance of horizontal pressure gradient and Coriolis force. This is called geostrophic balance.

On a nonrotating Earth, water would be accelerated by a horizontal pressure gradient and would flow from high to low pressure. On the rotating Earth, however, the Coriolis force deflects the motion, and the acceleration ceases only when the speed, U, of the current is just fast enough to produce a Coriolis force that can exactly balance the horizontal pressure-gradient force. This geostrophic balance is given as $dp/dx = \surd 2\omega \sin \theta$, and $dp/dy = -u2 \sin$, where dp/dx and dp/dy are the horizontal pressure gradient along the x-axis and y-axis, respectively, and u and v are the horizontal components of the velocity U along the x-axis and y-axis, respectively. From this balance it follows that the current direction must be perpendicular to the pressure gradient because the Coriolis force always acts perpendicular to the motion. In the Northern Hemisphere this direction is such that the high pressure is to the right when looking in current direction, while in the Southern Hemisphere it is to the left. This type of current is called a geostrophic current. The simple equation given above provides the basis for an indirect method of computing ocean currents. The relief of the sea surface also defines the streamlines (paths) of the geostrophic current at the surface relative to the deep reference level. The hills represent high pressure, and the valleys stand for low pressure. Clockwise rotation in the Northern Hemisphere with higher pressure in the centre of rotation is called anticyclonic motion. Counterclockwise rotation with lower pressure in its centre is cyclonic motion. In the Southern Hemisphere the sense of rotation is the opposite, because the effect of the Coriolis force has changed its sign of deflection.

Ekman Layer

The wind exerts stress on the ocean surface proportional to the square of the wind speed and in the direction of the wind, setting the surface water in motion. This motion extends to a depth of about

100 metres in what is called the Ekman layer, after the Swedish oceanographer V. Walfrid Ekman, who in 1902 deduced these results in a theoretical model constructed to help explain observations of wind drift in the Arctic. Within the oceanic Ekman layer the wind stress is balanced by the Coriolis force and frictional forces. The surface water is directed at an angle of 45° to the wind, to the right in the Northern Hemisphere and to the left in the Southern Hemisphere. With increasing depth in the boundary layer, the current speed is reduced, and the direction rotates farther away from the wind direction following a spiral form, becoming antiparallel to the surface flow at the base of the layer where the speed is 1/23 of the surface speed. This so-called Ekman spiral may be the exception rather than the rule, as the specific conditions are not often met, though deflection of a wind-driven surface current at somewhat smaller than 45° is observed when the wind field blows with a steady force and direction for the better part of a day. The average water particle within the Ekman layer moves at an angle of 90° to the wind; this movement is to the right of the wind direction in the Northern Hemisphere and to its left in the Southern Hemisphere. This phenomenon is called Ekman transport, and its effects are widely observed in the oceans.

Since the wind varies from place to place, so does the Ekman transport, forming convergence and divergence zones of surface water. A region of convergence forces surface water downward in a process called downwelling, while a region of divergence draws water from below into the surface Ekman layer in a process known as upwelling. Upwelling and downwelling also occur where the wind blows parallel to a coastline. The principal upwelling regions of the world are along the eastern boundary of the subtropical ocean waters, as, for example, the coastal region of Peru and northwestern Africa. Upwelling in these regions cools the surface water and brings nutrient-rich subsurface water into the sunlit layer of the ocean, resulting in a biologically productive region. Upwelling and high productivity also are found along divergence zones at the Equator and around Antarctica. The primary downwelling regions are in the subtropical ocean waters—e.g., the Sargasso Sea in the North Atlantic. Such areas are devoid of nutrients and are poor in marine life.

The vertical movements of ocean waters into or out of the base of the Ekman layer amount to less than 1 metre (about 3.3 feet) per day, but they are important since they extend the wind-driven effects into deeper waters. Within an upwelling region, the water column below the Ekman layer is drawn upward. This process, with conservation of angular momentum on the rotating Earth, induces the water column to drift toward the poles. Conversely, downwelling forces water into the water column below the Ekman layer, inducing drift toward the Equator. An additional consequence of upwelling and downwelling for stratified waters is to create a baroclinic field of mass. Surface water is less dense than deeper water. Ekman convergences have the effect of accumulating less dense surface water. This water floats above the surrounding water, forming a hill in sea level and driving an anticyclonic geostrophic current that extends well below the Ekman layer. Divergences do the opposite: they remove the less dense surface water, replacing it with denser, deeper water. This induces a depression in sea level with a cyclonic geostrophic current.

The ocean current pattern produced by the wind-induced Ekman transport is called the Sverdrup transport, after the Norwegian oceanographer H.U. Sverdrup, who formulated the basic theory in 1947. Several years later (1950) the American geophysicist and oceanographer Walter H. Munk and others expanded Sverdrup's work, explaining many of the major features of the wind-driven general circulation by using the mean climatological wind stress distribution at the sea surface as a driving force.

Two Types of Ocean Circulation

Ocean circulation derives its energy at the sea surface from two sources that define two circulation types: (1) wind-driven circulation forced by wind stress on the sea surface, inducing a momentum exchange, and (2) thermohaline circulation driven by the variations in water density imposed at the sea surface by exchange of ocean heat and water with the atmosphere, inducing a buoyancy exchange. These two circulation types are not fully independent, since the sea-air buoyancy and momentum exchange are dependent on wind speed. The wind-driven circulation is the more vigorous of the two and is configured as gyres that dominate an ocean region. The wind-driven circulation is strongest in the surface layer. The thermohaline circulation is more sluggish, with a typical speed of 1 cm (0.4 inch) per second, but this flow extends to the seafloor and forms circulation patterns that envelop the global ocean.

Wind-driven Circulation

Wind stress induces a circulation pattern that is similar for each ocean. In each case, the wind-driven circulation is divided into gyres that stretch across the entire ocean: subtropical gyres extend from the equatorial current system to the maximum westerlies in a wind field near 50° latitude, and subpolar gyres extend poleward of the maximum westerlies. The depth penetration of the wind-driven currents depends on the intensity of ocean stratification: in those regions of strong stratification, such as the tropics, the surface currents extend to a depth of less than 1,000 metres (about 3,300 feet), and within the low-stratification polar regions the wind-driven circulation reaches all the way to the seafloor.

Equatorial Currents

At the Equator the currents are for the most part directed toward the west, the North Equatorial Current in the Northern Hemisphere and the South Equatorial Current in the Southern Hemisphere. Near the thermal equator, where the warmest surface water is found, there occurs the eastward-flowing Equatorial Counter Current. This current is slightly north of the geographic Equator drawing the northern fringe of the South Equatorial Current to 5° N. The offset to the Northern Hemisphere matches a similar offset in the wind field. The east-to-west wind across the tropical ocean waters induces Ekman transport divergence at the Equator, which cools the surface water there.

At the geographic Equator a jetlike current is found just below the sea surface, flowing toward the east counter to the surface current. This is called the Equatorial Undercurrent. It attains speeds of more than 1 metre per second at a depth of nearly 100 metres. It is driven by higher sea level in the western margins of the tropical ocean, producing a pressure gradient, which in the absence of a horizontal Coriolis force drives a west-to-east current along the Equator. The wind field reverses the flow within the surface layer, inducing the South Equatorial Current.

Equatorial circulation undergoes variations following the irregular periods of roughly three to eight years of the Southern Oscillation (i.e., fluctuations of atmospheric pressure over the tropical Indo-Pacific region). Weakening of the east-to-west wind during a phase of the Southern Oscillation allows warm water in the western margin to slip back to the east by increasing the flow of the Equatorial Counter Current. Surface water temperatures and sea level decrease in the west and increase in the east. This event is called El Niño. The combined Southern Oscillation (ENSO) effect has received much attention because it is associated with global-scale climatic variability. In the tropical Indian Ocean the strong seasonal winds of the monsoons induce a similarly strong seasonal circulation pattern.

Subtropical Gyres

The subtropical gyres are anticyclonic circulation features. The Ekman transport within these gyres forces surface water to sink, giving rise to the subtropical convergence near 20°–30° latitude. The centre of the subtropical gyre is shifted to the west. This westward intensification of ocean currents was explained by the American meteorologist and oceanographer Henry M. Stommel as resulting from the fact that the horizontal Coriolis force increases with latitude. This causes the poleward-flowing western boundary current to be a jetlike current that attains speeds of 2 to 4 metres (6.5 to 13 feet) per second. This current transports the excess heat of the low latitudes to higher latitudes. The flow within the equatorward-flowing interior and eastern boundary of the subtropical gyres is quite different. It is more of a slow drift of cooler water that rarely exceeds 10 cm (about 4 inches) per second. Associated with these currents is coastal upwelling that results from offshore Ekman transport.

The strongest of the western boundary currents is the Gulf Stream in the North Atlantic Ocean. It carries about 30 million cubic metres (1 billion cubic feet) of ocean water per second through the Straits of Florida and roughly 80 million cubic metres (2.8 billion cubic feet) per second as it flows past Cape Hatteras off the coast of North Carolina, U.S. Responding to the large-scale wind field over the North Atlantic, the Gulf Stream separates from the continental margin at Cape Hatteras. After separation it forms waves or meanders that eventually generate many eddies of warm and cold water. The warm eddies, composed of thermocline water normally found south of the Gulf Stream, are injected into the waters of the continental slope off the coast of the northeastern United States. They drift to the southwest at rates of approximately 5 to 8 cm (about 2 to 3 inches) per second, and after a year they rejoin the Gulf Stream north of Cape Hatteras. Cold eddies of slope water are injected into the region south of the Gulf Stream and drift to the southwest. After roughly two years they reenter the Gulf Stream just north of the Antilles islands. The path that they follow defines a clockwise-flowing recirculation gyre seaward of the Gulf Stream.

Among the other western boundary currents, the Kuroshio of the North Pacific is perhaps the most like the Gulf Stream, having a similar transport and array of eddies. The Brazil Current and the East Australian Current are relatively weak. The Agulhas Current has a transport close to that of the Gulf Stream. It remains in contact with the margin of Africa around the southern rim of the continent. It then separates from the margin and curls back to the Indian Ocean in what is called the Agulhas Retroflection. Not all the water carried by the Agulhas Current returns to the east; about 10 to 20 percent is injected into the South Atlantic Ocean as large eddies that slowly migrate across it.

Subpolar Gyres

The subpolar gyres are cyclonic circulation features. The Ekman transport within these features forces upwelling and surface water divergence. In the North Atlantic the subpolar gyre consists of the North Atlantic Current at its equatorward side and the Norwegian Current that carries relatively warm water northward along the coast of Norway. The heat released from the Norwegian Current into the atmosphere maintains a moderate climate in northern Europe. Along the east coast of Greenland is the southward-flowing cold East Greenland Current. It loops around the southern tip of Greenland and continues flowing into the Labrador Sea. The southward flow that continues off the coast of Canada is called the Labrador Current. This current separates for the most part from the coast near Newfoundland to complete the subpolar gyre of the North Atlantic. Some of the cold water of the Labrador Current, however, extends farther south.

In the North Pacific the subpolar gyre is composed of the northward-flowing Alaska Current, the Aleutian Current, and the southward-flowing cold Oyashio Current. The North Pacific Current forms the separation between the subpolar and subtropical gyres of the North Pacific.

In the Southern Hemisphere the subpolar gyres are less defined. Large cyclonic flowing gyres lie poleward of the Antarctic Circumpolar Current and can be considered counterparts to the Northern Hemispheric subpolar gyres. The best-formed is the Weddell Gyre of the South Atlantic sector of the Southern Ocean. The Antarctic coastal current flows toward the west. The northward-flowing current off the east coast of the Antarctic Peninsula carries cold Antarctic coastal water into the circumpolar belt. Another cyclonic gyre occurs north of the Ross Sea.

Antarctic Circumpolar Current

The Southern Ocean links the major oceans by a deep circumpolar belt in the 50°–60° S range. In this belt flows the Antarctic Circumpolar Current from west to east, encircling the globe at high latitudes. It transports 125 million cubic metres (4.4 billion cubic feet) of seawater per second over a path of about 24,000 km (about 14,900 miles) and is the most important factor in diminishing the differences between oceans. The Antarctic Circumpolar Current is not a well-defined single-axis current but rather consists of a series of individual currents separated by frontal zones. It reaches the seafloor and is guided along its course by the irregular bottom topography. Large meanders and eddies develop in the current as it flows. These features induce poleward transfer of heat, which may be significant in balancing the oceanic heat loss to the atmosphere above the Antarctic region farther south.

Thermohaline Circulation

The general circulation of the oceans consists primarily of the wind-driven currents. These, however, are superimposed on the much more sluggish circulation driven by horizontal differences in temperature and salinity—namely, the thermohaline circulation. The thermohaline circulation reaches down to the seafloor and is often referred to as the deep, or abyssal, ocean circulation. Measuring seawater temperature and salinity distribution is the chief method of studying the deep-flow patterns. Other properties also are examined; for example, the concentrations of oxygen, carbon-14, and such synthetically produced compounds as chlorofluorocarbons are measured to obtain resident times and spreading rates of deep water.

In the process it transports heat, which influences regional climate patterns. The density of seawater is determined by the temperature and salinity of a volume of seawater at a particular location. The difference in density between one location and another drives the thermohaline circulation.

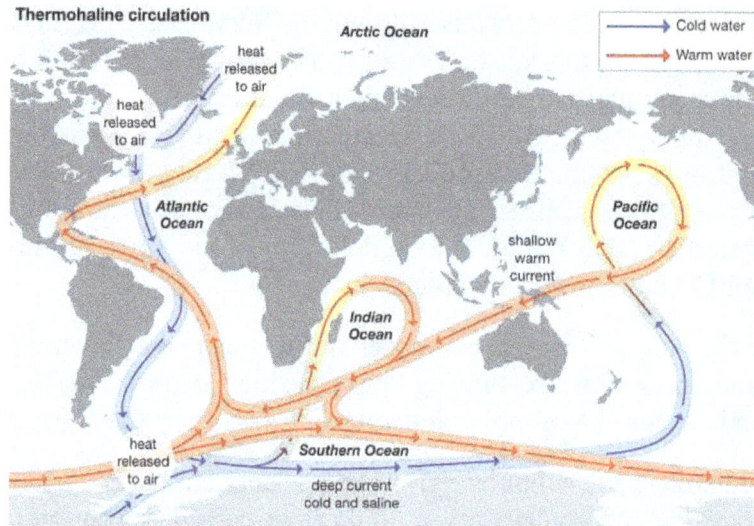

Thermohaline circulation transports and mixes the water of the oceans.

In some areas of the ocean, generally during the winter season, cooling or net evaporation causes surface water to become dense enough to sink. Convection penetrates to a level where the density of the sinking water matches that of the surrounding water. It then spreads slowly into the rest of the ocean. Other water must replace the surface water that sinks. This sets up the thermohaline circulation. The basic thermohaline circulation is one of sinking of cold water in the polar regions, chiefly in the northern North Atlantic and near Antarctica. These dense water masses spread into the full extent of the ocean and gradually up well to feed a slow return flow to the sinking regions. A theory for the thermohaline circulation pattern was proposed by Stommel and Arnold Arons in 1960.

In the Northern Hemisphere the primary region of deep water formation is the North Atlantic; minor amounts of deep water are formed in the Red Sea and Persian Gulf. A variety of water types contribute to the so-called North Atlantic Deep Water. Each one of them differs, though they share a common attribute of being relatively warm (greater than 2 °C) and salty (greater than 34.9 parts per thousand) compared with the other major producer of deep and bottom water, the Southern Ocean (0° C and 34.7 parts per thousand). North Atlantic Deep Water is primarily formed in the Greenland and Norwegian seas, where cooling of the salty water introduced by the Norwegian Current induces sinking. This water spills over the rim of the ridge that stretches from Greenland to Scotland, extending to the seafloor to the south as a convective plume. It then flows southward, pressed against the western edge of the North Atlantic. Additional deep water is formed in the Labrador Sea. This water, somewhat less dense than the overflow water from the Greenland and Norwegian seas, has been observed sinking to a depth of 3,000 metres (about 9,800 feet) within convective features referred to as chimneys. Vertical velocities as high as 10 cm per second have been observed within these convective features. A third variety of North Atlantic Deep Water is derived from net evaporation within the Mediterranean Sea. This draws surface water into the Mediterranean through the Strait of Gibraltar. The mass of salty water formed within the Mediterranean exits as a deeper stream. It descends to depths of approximately 1,000 metres in the North Atlantic Ocean, forming the uppermost layer of North Atlantic Deep Water. The outflow in the Strait of Gibraltar reaches as high as 2 metres per second, but its total transport amounts to only 5 percent of the total North Atlantic Deep Water formed. The outflow of the Mediterranean plays a significant role in boosting the salinity of North Atlantic Deep Water.

The blend of North Atlantic Deep Water, with a total formation rate of 15 to 20 million cubic metres (530 to 706 million cubic feet) per second, quickly ventilates the Atlantic Ocean, resulting in a residence time of less than 200 years. The deep water spreads away from its source along the western side of the Atlantic Ocean and, on reaching the Antarctic Circumpolar Current, spreads into the Indian and Pacific oceans. The sinking of North Atlantic Deep Water is compensated for by the slow upwelling of deep water, mainly in the Southern Ocean, to replenish the upper stratum of water that has descended as North Atlantic Deep Water. North Atlantic Deep Water exported to the other oceans must be balanced by the inflow of upper-layer water into the Atlantic. Some water returns as cold, low-salinity Pacific water through the Drake Passage in the form of what is known as Antarctic Intermediate Water, and some returns as warm salty thermocline water from the Indian Ocean around the southern rim of Africa.

Remnants of North Atlantic Deep Water mix with Southern Ocean water to spread along the sea-floor into the North Pacific Ocean. Here it upwells to a level of 2,000–3,000 metres (about 6,500–9,800 feet) and returns to the south lower in salinity and oxygen but higher in nutrient concentrations as North Pacific Deep Water. This North Pacific Deep Water is eventually swept eastward with the Antarctic Circumpolar Current. Modification of deep water in the North Pacific is the direct consequence of vertical mixing, which carries into the deep ocean the low salinity properties of North Pacific Intermediate Water. The latter is formed in the northwestern Pacific Ocean. Because of the immenseness of the North Pacific and the extremely long residence time (more than 500 years) of the water, enormous quantities of North Pacific Deep Water can be produced by vertical mixing.

Considerable volumes of cold water generally of low salinity are formed in the Southern Ocean. Such water masses spread into the interior of the global ocean and to a large extent are responsible for the anomalous cold, low-salinity state of the modern oceans. The circumstances leading to this role for the Southern Ocean are related to the existence of a deep-ocean circumpolar belt around Antarctica that was established some 25 million years ago by the shifting lithospheric plates which make up Earth's surface. This belt establishes the Antarctic Circumpolar Current, which isolates Antarctica from the warm surface waters of the subtropics. The Antarctic Circumpolar Current does not completely sever contact with the lower latitudes. The Southern Ocean does have access to the waters of the north, but through deep- and bottom-water pathways. The basic dynamics of the Antarctic Circumpolar Current lift dense deep water occurring north of the current to the ocean surface south of it. Once exposed to the cold Antarctic air masses, the upwelling deep water is converted to the cold Antarctic Bottom Water and Antarctic Intermediate Water. The southward and upwelling deep water, which carries heat injected into the deep ocean by processes farther north, is balanced by the northward spread of cooler, fresher, oxygenated water masses of the Southern Ocean. It is estimated that the overturning rate of water south of the Antarctic Circumpolar Current amounts to 35 to 45 million cubic metres (1.2 to 1.6 billion cubic feet) per second, most of which becomes Antarctic Bottom Water.

The primary site of Antarctic Bottom Water formation is within the continental margins of the Weddell Sea, though some is produced in other coastal regions, such as the Ross Sea. Also, there is evidence of deep convective overturning farther offshore. Antarctic Bottom Water, formed at a rate of 30 million cubic metres per second, slips below the Antarctic Circumpolar Current and spreads to regions well north of the Equator. Slowly upwelling and modified by mixing with less dense water, it returns to the Southern Ocean as deep water.

The remaining upwelling of deep water spreads near the surface to the north, where it forms Antarctic Intermediate Water within the Antarctic Circumpolar Current zone and spreads along the base of the thermoclines farther north. This water mass form a sheet of low-salinity water that demarcates the lower boundary of the subtropical thermocline. It upwells into the thermocline, partly compensating for the sinking of North Atlantic Deep Water.

Ocean Current Energy

Marine currents can carry large amounts of energy, largely driven by the tides, which are a consequence of the gravitational effects of the planetary motion of the Earth, the Moon and the Sun. Augmented flow velocities can be found where the underwater topography (bathymetry) in straits between islands and the mainland or in shallows around headlands plays a major role in enhancing the flow velocities, resulting in appreciable kinetic energy. The sun acts as the primary driving force, causing winds and temperature differences. Because there are only small fluctuations in current speed and stream location with minimal changes in direction, ocean currents may be suitable locations for deploying energy extraction devices such as turbines. Other effects such as regional differences in temperature and salinity and the Coriolis effect due to the rotation of the earth are also major influences. The kinetic energy of marine currents can be converted in much the same way that a wind turbine extracts energy from the wind, using various types of open-flow rotors.

Energy Potential

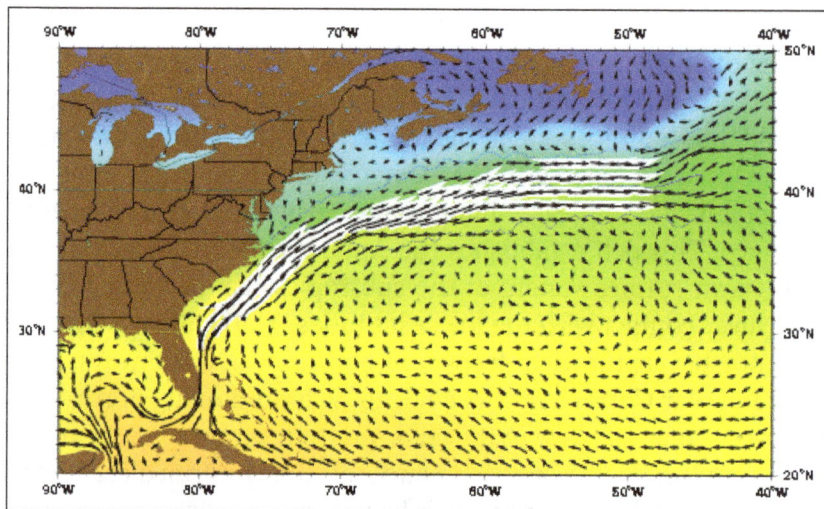

Vector Diagram of current flow along the east coast.

The total worldwide power in ocean currents has been estimated to be about 5,000 GW, with power densities of up to 15 kW/m². The relatively constant extractable energy density near the surface of the Florida Straits Current is about 1 kW/m² of flow area. It has been estimated that capturing just 1/1,000th of the available energy from the Gulf Stream, which has 21,000 times more energy than Niagara Falls in a flow of water that is 50 times the total flow of all the world's freshwater rivers, would supply Florida with 35% of its electrical needs. The image to the right illustrates the high density of flow along the coast, note the high velocity white northward flow, perfect for extraction of ocean current energy. Countries that are interested in and pursuing the application of ocean current energy technologies include the European Union, Japan, and China.

The potential of electric power generation from marine tidal currents is enormous. There are several factors that make electricity generation from marine currents very appealing when compared to other renewables:

- The high load factors resulting from the fluid properties. The predictability of the resource, so that, unlike most of other renewables, the future availability of energy can be known and planned for.

- The potentially large resource that can be exploited with little environmental impact, thereby offering one of the least damaging methods for large-scale electricity generation.

- The feasibility of marine-current power installations to provide also base grid power, especially if two or more separate arrays with offset peak-flow periods are interconnected.

Technologies for Marine-current-power Generation

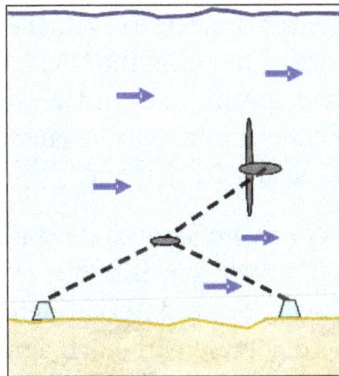

Windpower inspired axial flow turbine
used for marine power generation.

There are several types of open-flow devices that can be used in marine-current-power applications; many of them are modern descendants of the waterwheel or similar. However, the more technically sophisticated designs, derived from wind-power rotors, are the most likely to achieve enough cost-effectiveness and reliability to be practical in a massive marine-current-power future scenario. Even though there is no generally accepted term for these open-flow hydro-turbines, some sources refer to them as water-current turbines. There are two main types of water current turbines that might be considered: axial-flow horizontal-axis propellers (with both variable-pitch or fixed-pitch), and cross-flow Darrieus rotors.

An example of helical-style Darrieus rotor turbines installed to capture energy
from river currents, this technology is also employed for marine uses.

Both rotor types may be combined with any of the three main methods for supporting water-current turbines: floating moored systems, sea-bed mounted systems, and intermediate systems. Sea-bed-mounted monopile structures constitute the first-generation marine current power systems. They have the advantage of using existing (and reliable) engineering, but they are limited to relatively shallow waters (about 20 to 40 m depth).

Applicatons

The possible use of marine currents as an energy resource began to draw attention in the mid-1970s after the first oil crisis. In 1974 several conceptual designs were presented at the MacArthur Workshop on Energy, and in 1976 the British General Electric Company. undertook a partially government-funded study which concluded that marine current power deserved more detailed research. Soon after, the ITD-Group in UK implemented a research program involving a year performance testing of a 3-m hydroDarrieus rotor deployed at Juba on the White Nile.

The 1980s saw a number of small research projects to evaluate marine current power systems. The main countries where studies were carried out were the UK, Canada, and Japan. In 1992–1993 the Tidal Stream Energy Review identified specific sites in UK waters with suitable current speed to generate up to 58 TWh/year. It confirmed a total marine current power resource capable theoretically of meeting some 19% of the UK electricity demand.

In 1994–1995 the EU-JOULE CENEX project identified over 100 European sites ranging from 2 to 200 km² of sea-bed area, many with power densities above 10 MW/km². Both the UK Government and the EU have committed themselves to internationally negotiated agreements designed to combat global warming. In order to comply with such agreements, an increase in large-scale electricity generation from renewable resources will be required. Marine currents have the potential to supply a substantial share of future EU electricity needs. The study of 106 possible sites for tidal turbines in the EU showed a total potential for power generation of about 50 TWh/year. If this resource is to be successfully utilized, the technology required could form the basis of a major new industry to produce clean power for the 21st century.

Contemporary applications of these technologies can be found here: Since the effects of tides on ocean currents are so large, and their flow patterns are quite reliable, many ocean current energy extraction plants are placed in areas of high tidal flow rates.

Research on marine current power is conducted at, among others, Uppsala University in Sweden, where a test unit with a straight-bladed Darrieus type turbine has been constructed and placed in the Dal river in Sweden.

Environmental Effects

Ocean currents are instrumental in determining the climate in many regions around the world. While little is known about the effects of removing ocean current energy, the impacts of removing current energy on the farfield environment may be a significant environmental concern. The typical turbine issues with blade strike, entanglement of marine organisms, and acoustic effects still exists; however, these may be magnified due to the presence of more diverse populations of marine organisms using ocean currents for migration purposes. Locations can be further offshore and therefore require longer power cables that could affect the marine environment with electromagnetic output.

Boundary Current

Boundary currents are ocean currents with dynamics determined by the presence of a coastline, and fall into two distinct categories: western boundary currents and eastern boundary currents.

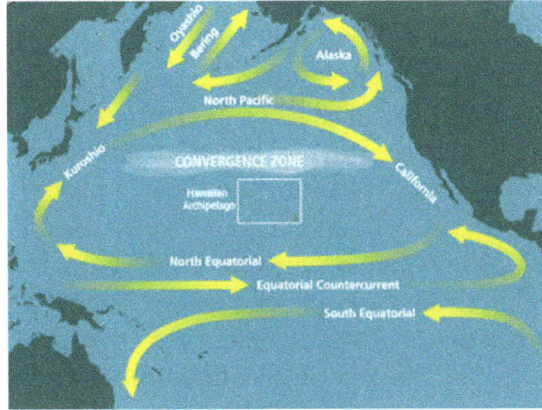

The main ocean currents involved with the North Pacific Gyre.

Eastern Boundary Currents

Eastern boundary currents are relatively shallow, broad and slow-flowing. They are found on the eastern side of oceanic basins (adjacent to the western coasts of continents). Subtropical eastern boundary currents flow equatorward, transporting cold water from higher latitudes to lower latitudes; examples include the Benguela Current, the Canary Current, the Humboldt Current, and the California Current. Coastal upwelling often brings nutrient-rich water into eastern boundary current regions, making them productive areas of the ocean.

Western Boundary Currents

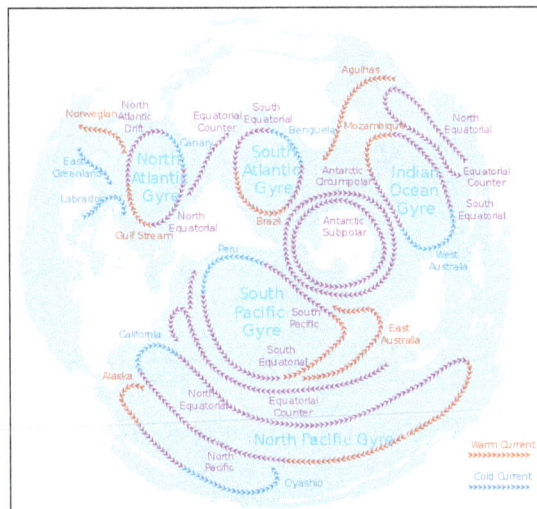

The world's largest ocean gyres.

Western boundary currents are warm, deep, narrow, and fast flowing currents that form on the west side of ocean basins due to western intensification. They carry warm water from the tropics poleward. Examples include the Gulf Stream, the Agulhas Current, and the Kuroshio.

Western Intensification

Western intensification is the intensification of the western arm of an oceanic current, particularly a large gyre in an ocean basin. The trade winds blow westward in the tropics, and the westerlies blow eastward at mid-latitudes. This wind pattern applies a stress to the subtropical ocean surface with negative curl in the northern hemisphere and a positive curl in the southern hemisphere. The resulting Sverdrup transport is equatorward in both cases. Because of conservation of mass and potential vorticity conservation, that transport is balanced by a narrow, intense poleward current, which flows along the western boundary of the ocean basin, allowing the vorticity introduced by coastal friction to balance the vorticity input of the wind. Western intensification also occurs in the polar gyres, where the sign of the wind stress curl and the direction of the resulting currents are reversed. It is because of western intensification that the currents on the western boundary of a basin (such as the Gulf Stream, a current on the western side of the Atlantic Ocean) are stronger than those on the eastern boundary (such as the California Current, on the eastern side of the Pacific Ocean). Western intensification was first explained by the American oceanographer Henry Stommel.

In 1948, Henry Stommel published a paper in *Transactions, American Geophysical Union* titled "The Westward Intensification of Wind-Driven Ocean Currents", in which he used a simple, homogeneous, rectangular ocean model to examine the streamlines and surface height contours for an ocean at a non-rotating frame, an ocean characterized by a constant Coriolis parameter and finally, a real-case ocean basin with a latitudinally-varying Coriolis parameter. In this simple, modeling setting, the principal factors that were accounted for influencing the oceanic circulation were surface wind stress, bottom friction, a variable surface height leading to horizontal pressure gradients, and finally, the Coriolis effect.

In his simplified model, he assumed an ocean of constant density and depth D + h in the presence of ocean currents; he also introduced a linearized, frictional term to account for the dissipative effects that prevent the real ocean from accelerating. He starts, thus, from the steady-state momentum and continuity equations:

$$f(D+h)v - F\cos\left(\frac{\pi y}{b}\right) - Ru - g(D+h)\frac{\partial h}{\partial x} = 0$$

$$-f(D+h)u - Rv - g(D+h)\frac{\partial h}{\partial y} = 0$$

$$\frac{\partial[(D+h)u]}{\partial x} + \frac{\partial[(D+h)v]}{\partial y} = 0$$

Here f is the strength of the Coriolis force, R is the bottom-friction coeffecient, g is gravity, and $-F\cos\left(\frac{\pi y}{b}\right)$ is the wind forcing. The wind is blowing towards the west at y = 0 and towards the east at y = b.

Acting on $\frac{\partial}{\partial y}$ and $\frac{\partial}{\partial x}$, subtracting, then gives,

$$v(D+h)\left(\frac{\partial f}{\partial y}\right) + \frac{\pi F}{b}\sin\left(\frac{\pi y}{b}\right) + R\left(\frac{\partial v}{\partial x} - \frac{\partial u}{\partial y}\right) = 0$$

If we introduce a Stream function ψ and linearize by assuming that $D \gg h$, equation just above reduces to:

$$\nabla^2 \psi + \alpha \left(\frac{\partial \psi}{\partial x} \right) = \gamma \sin \left(\frac{\pi y}{b} \right)$$

here

$$\alpha = \left(\frac{D}{R} \right) \left(\frac{\partial f}{\partial y} \right)$$

and

$$\gamma = \frac{\pi F}{Rb}$$

The solutions of $\nabla^2 \psi + \alpha \left(\frac{\partial \psi}{\partial x} \right) = \gamma \sin \left(\frac{\pi y}{b} \right)$ with boundary condition that ψ be constant on the coastlines, and for different values of α, emphasize the role of the variation of the Coriolis parameter with latitude in inciting the strengthening of western boundary currents. Such currents are observed to be much faster, deeper, narrower and warmer than their eastern counterparts.

For a non-rotating state (zero Coriolis parameter) as well as an ocean state at which the Coriolis parameter is a constant, the ocean circulation does not demonstrate any preference toward intensification/acceleration near the western boundary. The streamlines exhibit a symmetric behavior in all directions, with the height contours demonstrating a nearly parallel relation to the streamlines, in the case of the homogeneously rotating ocean. Finally, for the case of interest - the one in which the Coriolis force is latitudinally variant - a distinct tendency for an asymmetrical streamline diagram is noted, with an observed, intense clustering toward the western part of the modeled ocean. A nice set of figures depicting the distribution of streamlines and height contours for the cases of a uniformly-rotating ocean and an ocean where the Coriolis force is linearly dependent on latitude can be found in Stommel's 1948 paper.

Sverdrup Balance and Physics of Western Intensification

The physics of western intensification can be understood through a mechanism that helps maintain the vortex balance along an ocean gyre. Harald Sverdrup was the first one, preceding Henry Stommel, to attempt to explain the mid-ocean vorticity balance by looking at the relationship between surface wind forcings and the mass transport within the upper ocean layer. He assumed a geostrophic interior flow, while neglecting any frictional or viscosity effects and presuming that the circulation vanishes at some depth in the ocean. This prohibited the application of his theory to the western boundary currents, since some form of dissipative effect would be later shown to be necessary to predict a closed circulation for an entire ocean basin and to counteract the wind-driven flow.

Sverdrup introduced a potential vorticity argument to connect the net, interior flow of the oceans to the surface wind stress and the incited planetary vorticity perturbations. For instance, Ekman convergence in the sub-tropics (related to the existence of the trade winds in the tropics and the westerlies in the mid-latitudes) was suggested to lead to a downward vertical velocity and therefore, a squashing of the water columns, which subsequently forces the ocean gyre to spin more slowly (via angular momentum conservation). This is accomplished via a decrease in planetary vorticity (since relative vorticity variations are not significant in large ocean circulations), a phenomenon attainable through an equator-wardly directed, interior flow that characterizes the subtropical gyre. The opposite is applicable when Ekman divergence is induced, leading to Ekman absorption (suction) and a subsequent, water column stretching and poleward return flow, a characteristic of sub-polar gyres.

This return flow, as shown by Stommel, occurs in a meridional current, concentrated near the western boundary of an ocean basin. To balance the vorticity source induced by the wind stress forcing, Stommel introduced a linear frictional term in the Sverdrup equation, functioning as the vorticity sink. This bottom ocean, frictional drag on the horizontal flow allowed Stommel to theoretically predict a closed, basin-wide circulation, while demonstrating the west-ward intensification of wind-driven gyres and its attribution to the Coriolis variation with latitude (beta effect). Walter Munk further implemented Stommel's theory of western intensification by using a more realistic frictional term, while emphasizing "the lateral dissipation of eddy energy." In this way, not only did he reproduce Stommel's results, recreating thus the circulation of a western boundary current of an ocean gyre resembling the Gulf stream, but he also showed that sub-polar gyres should develop northward of the subtropical ones, spinning in the opposite direction.

Geostrophic Current

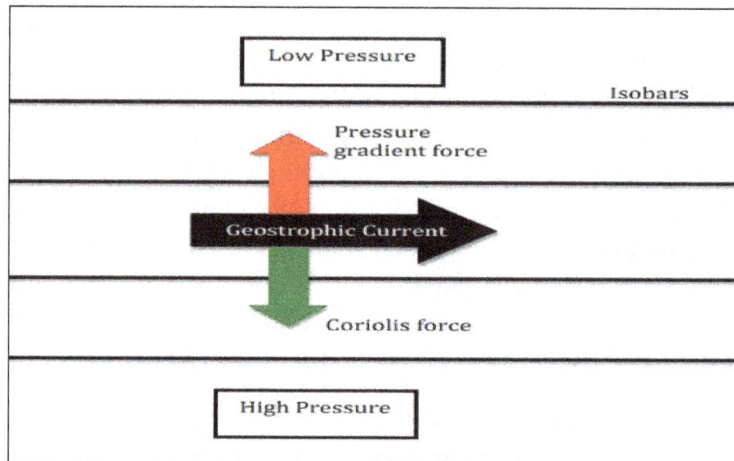

A geostrophic current is an oceanic current in which the pressure gradient force is balanced by the Coriolis effect. The direction of geostrophic flow is parallel to the isobars, with the high pressure to the right of the flow in the Northern Hemisphere, and the high pressure to the left in the Southern Hemisphere. This concept is familiar from weather maps, whose isobars show the direction of geostrophic flow in the atmosphere. Geostrophic flow may be either barotropic or baroclinic. A geostrophic current may also be thought of as a rotating shallow water wave with a frequency of zero. The principle of geostrophy is useful to oceanographers because it allows them to infer ocean currents from measurements of the sea surface height (by combined satellite altimetry and gravimetry) or from vertical profiles of seawater density taken by ships or autonomous buoys. The major currents of the world's oceans, such as the Gulf Stream, the Kuroshio Current, the Agulhas

Current, and the Antarctic Circumpolar Current, are all approximately in geostrophic balance and are examples of geostrophic currents.

Simple Explanation

Sea water naturally tends to move from a region of high pressure (or high sea level) to a region of low pressure (or low sea level). The force pushing the water towards the low pressure region is called the pressure gradient force. In a geostrophic flow, instead of water moving from a region of high pressure (or high sea level) to a region of low pressure (or low sea level), it moves along the lines of equal pressure (isobars). This occurs because the Earth is rotating. The rotation of the earth results in a "force" being felt by the water moving from the high to the low, known as Coriolis force. The Coriolis force acts at right angles to the flow, and when it balances the pressure gradient force, the resulting flow is known as geostrophic.

As stated above, the direction of flow is with the high pressure to the right of the flow in the Northern Hemisphere, and the high pressure to the left in the Southern Hemisphere. The direction of the flow depends on the hemisphere, because the direction of the Coriolis force is opposite in the different hemispheres.

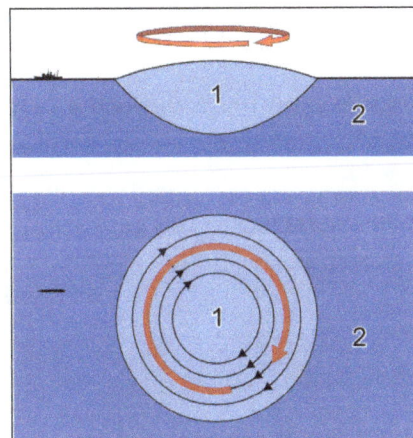

A northern-hemisphere gyre in geostrophic balance.

Paler water is less dense than dark water, but more dense than air; the outwards pressure gradient is balanced by the 90 degrees-right-of-flow coriolis force. The structure will eventually dissipate due to friction and mixing of water properties.

Formulation

The geostrophic equations are a simplified form of the Navier–Stokes equations in a rotating reference frame. In particular, it is assumed that there is no acceleration (steady-state), that there is no viscosity, and that the pressure is hydrostatic. The resulting balance:

$$fv = \frac{1}{\rho}\frac{\partial p}{\partial x}$$

$$fu = -\frac{1}{\rho}\frac{\partial p}{\partial y}$$

Where, f is the Coriolis parameter, ρ is the density, p is the pressure and u, v are the velocities in the x, y -directions respectively. One special property of the geostrophic equations, is that they satisfy the steady-state version of the continuity equation. That is:

$$\frac{\partial u}{\partial x}+\frac{\partial v}{\partial y}=0$$

Rotating Waves of Zero Frequency

The equations governing a linear, rotating shallow water wave are:

$$\frac{\partial u}{\partial t}-fv=-\frac{1}{\rho}\frac{\partial p}{\partial x}$$

$$\frac{\partial v}{\partial t}+fu=-\frac{1}{\rho}\frac{\partial p}{\partial y}$$

The assumption of steady-state made above (no acceleration) is:

$$\frac{\partial u}{\partial t}=\frac{\partial v}{\partial t}=0$$

Alternatively, we can assume a wave-like, periodic, dependence in time:

$$u \propto v \propto e^{i\omega t}$$

In this case, if we set $\omega = 0$, we have reverted to the geostrophic equations above. Thus a geostrophic current can be thought of as a rotating shallow water wave with a frequency of zero.

Ocean Tide

The Bay of Fundy at high tide.   The Bay of Fundy at low tide.

Tides are the cyclic rising and falling of the Earth's ocean surface caused by the tidal forces of the Moon and Sun acting on the oceans. Tides cause changes in the depth of the marine and estuarine water bodies and produce oscillating currents known as tidal streams, making prediction of tides important for coastal navigation. The strip of seashore that is submerged at high tide and exposed at low tide, called the intertidal zone, is an important ecological product of ocean tides.

The changing tide produced at a given location is the result of several factors, including the changing positions of the Moon and Sun relative to the Earth, the effects of Earth's rotation, and local water depth. Sea level measured by coastal tide gauges may also be strongly affected by wind. More generally, tidal phenomena can occur in other systems besides the ocean, whenever a gravitational field that varies in time and space is present.

Cause of Ocean Tide

High tides and low tides are caused by the moon. The moon's gravitational pull generates something called the tidal force. The tidal force causes Earth—and its water—to bulge out on the side closest to the moon and the side farthest from the moon. These bulges of water are high tides.

As the Earth rotates, your region of Earth passes through both of these bulges each day. When you're in one of the bulges, you experience a high tide. When you're not in one of the bulges, you experience a low tide. This cycle of two high tides and two low tides occurs most days on most of the coastlines of the world.

Tidal Range Variation: Springs and Neaps

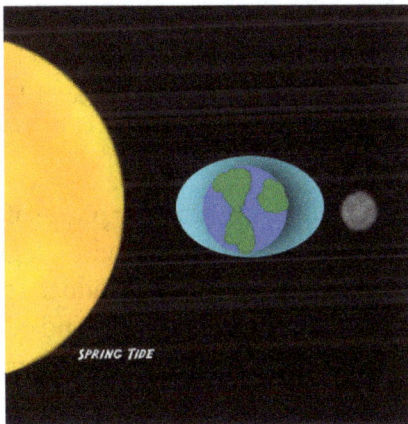

An artist's conception of spring tide.   An artist's conception of neap tide.

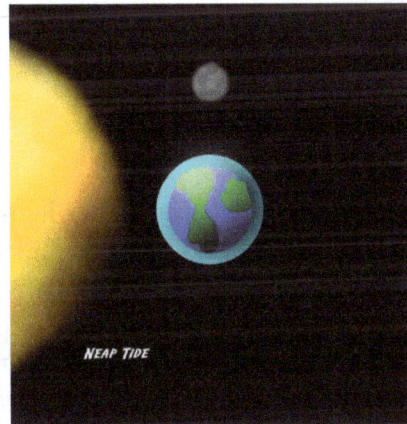

The semidiurnal tidal range (the difference in height between high and low tides over about a half day) varies in a two-week or fortnightly cycle. Around new and full moon when the Sun, Moon and Earth form a line (a condition known as syzygy), the tidal forces due to the Sun reinforce those of the Moon. The tide's range is then maximum: this is called the spring tide, or just springs and is derived not from the season of spring but rather from the verb meaning "to jump" or "to leap up." When the Moon is at first quarter or third quarter, the Sun and Moon are separated by 90° when viewed from the earth, and the forces due to the Sun partially cancel those of the Moon. At these points in the lunar cycle, the tide's range is minimum, this is called the neap tide, or neaps. Spring tides result in high waters that are higher than average, low waters that are lower than average, slack water time that is shorter than average and stronger tidal currents than average. Neaps result in less extreme tidal conditions. There is about a seven day interval between springs and neaps.

The changing distance of the Moon from the Earth also affects tide heights. When the Moon is at perigee the range is increased and when it is at apogee the range is reduced. Every 7½ lunations, perigee and (alternately) either a new or full moon coincide causing *perigean tides* with the largest *tidal range,* and if a storm happens to be moving onshore at this time, the consequences (in the form of property damage, etc.) can be especially severe.

Tidal Phase and Amplitude

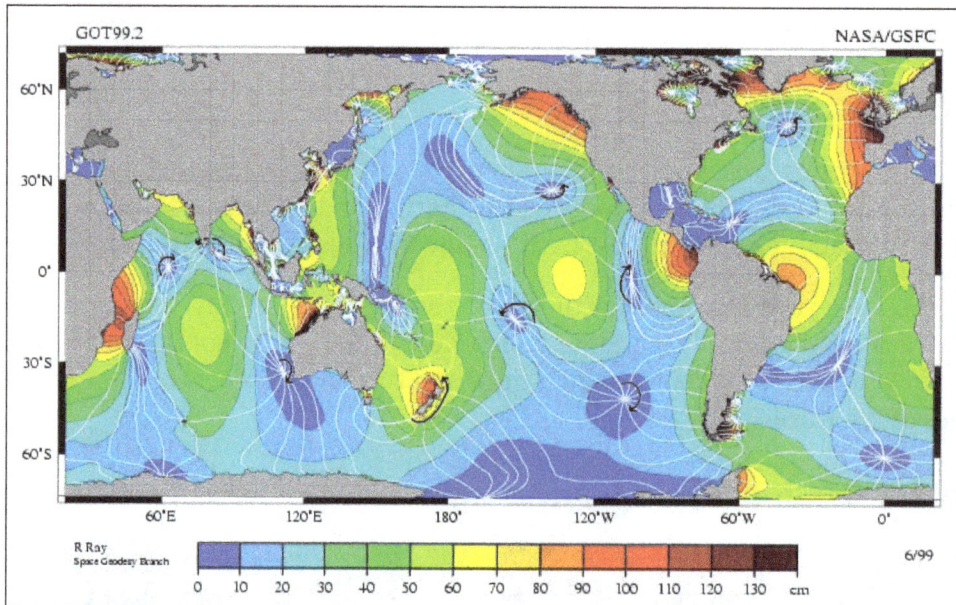

The M_2 tidal constituent: Amplitude is indicated by color, and the white lines are cotidal differing by 1 hr. The curved arcs around the amphidromic points show the direction of the tides, each indicating a synchronized 6 hour period.

Because the M2 tidal constituent dominates in most locations, the stage or phase of a tide, denoted by the time in hours after high tide, is a useful concept. It is also measured in degrees, with 360° per tidal cycle. Lines of constant tidal phase are called cotidal lines. High tide is reached simultaneously along the cotidal lines extending from the coast out into the ocean, and cotidal lines (and hence tidal phases) advance along the coast. If one thinks of the ocean as a circular basin enclosed by a coastline, the cotidal lines point radially inward and must eventually meet at a common point, the amphidromic point. An amphidromic point is at once cotidal with high and low tides, which is satisfied by zero tidal motion. (The rare exception occurs when the tide circles around an island, as it does around New Zealand.) Indeed tidal motion generally lessens moving away from the continental coasts, so that crossing the cotidal lines are contours of constant amplitude (half of the distance between high and low tide) which decrease to zero at the amphidromic point. For a 12 hour semidiurnal tide the amphidromic point behaves roughly like a clock face, with the hour hand pointing in the direction of the high tide cotidal line, which is directly opposite the low tide cotidal line. High tide rotates about once every 12 hours in the direction of rising cotidal lines, and away from ebbing cotidal lines. The difference of cotidal phase from the phase of a reference tide is the epoch.

The shape of the shoreline and the ocean floor change the way that tides propagate, so there is no simple, general rule for predicting the time of high tide from the position of the Moon in the sky.

Coastal characteristics such as underwater topography and coastline shape mean that individual location characteristics need to be taken into consideration when forecasting tides; high water time may differ from that suggested by a model such as the one above due to the effects of coastal morphology on tidal flow.

Tidal Physics

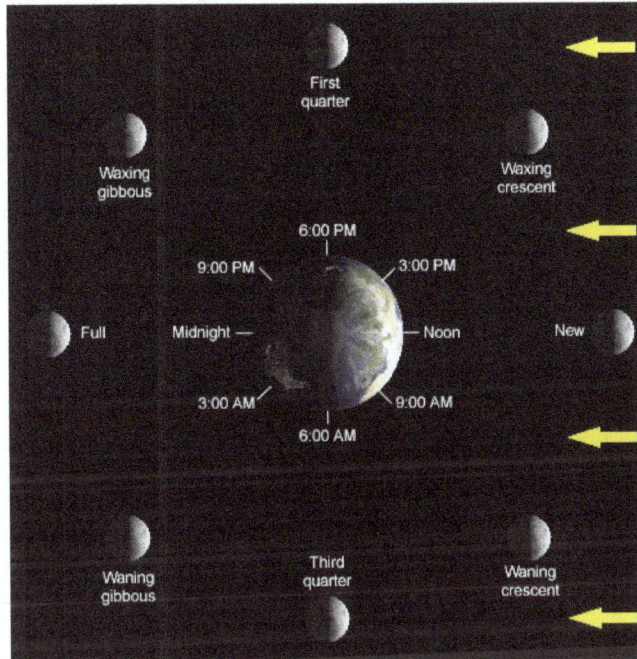

The Earth and Moon, looking at the North Pole.

Isaac Newton laid the foundations for the mathematical explanation of tides in the Philosophiae Naturalis Principia Mathematica. In 1740, the Académie Royale des Sciences in Paris offered a prize for the best theoretical essay on tides. Daniel Bernoulli, Antoine Cavalleri, Leonhard Euler, and Colin Maclaurin shared the prize. Maclaurin used Newton's theory to show that a smooth sphere covered by a sufficiently deep ocean under the tidal force of a single deforming body is a prolate spheroid with major axis directed toward the deforming body. Maclaurin was also the first to write about the Earth's rotational effects on motion. Euler realized that the horizontal component of the tidal force (more than the vertical) drives the tide. In 1744 D'Alembert studied tidal equations for the atmosphere which did not include rotation. The first major theoretical formulation for water tides was made by Pierre-Simon Laplace, who formulated a system of partial differential equations relating the horizontal flow to the surface height of the ocean. The Laplace tidal equations are still in use today. William Thomson rewrote Laplace's equations in terms of vorticity which allowed for solutions describing tidally driven coastally trapped waves, which are known as Kelvin waves.

Tidal Forces

The tidal force produced by a massive object (Moon, hereafter) on a small particle located on or in an extensive body (Earth, hereafter) is the vector difference between the gravitational force exerted by the Moon on the particle, and the gravitational force that would be exerted on the particle if it were located at the center of mass of the Earth. Thus, the tidal force depends not on the strength of the gravitational field of the Moon, but on its gradient. The gravitational force exerted on the Earth by the Sun is on average 179 times stronger than that exerted on the Earth by the Moon, but because the Sun is on average 389 times farther from the Earth, the gradient of its field is weaker. The tidal force produced by the Sun is therefore only 46% as large as that produced by the Moon.

Tidal forces can also be analyzed from the point of view of a reference frame that translates with the center of mass of the Earth. Consider the tide due to the Moon (the Sun is similar). First observe that the Earth and Moon rotate around a common orbital center of mass, as determined by their relative masses. The orbital center of mass is 3/4 of the way from the Earth's center to its surface. The second observation is that the Earth's centripetal motion is the averaged response of the entire Earth to the Moon's gravity and is exactly the correct motion to balance the Moon's gravity only at the center of the Earth; but every part of the Earth moves along with the center of mass and all parts have the same centripetal motion, since the Earth is rigid. On the other hand each point of the Earth experiences the Moon's radially decreasing gravity differently; the near parts of the Earth are more strongly attracted than is compensated by the centripetal motion and experience a net tidal force toward the Moon; the far parts have more centripetal motion than is necessary for the reduced attraction, and thus feel a net force away from the Moon. Finally only the horizontal components of the tidal forces actually contribute tidal acceleration to the water particles since there is small resistance. The actual tidal force on a particle is only about a ten millionth of the force caused by the Earth's gravity.

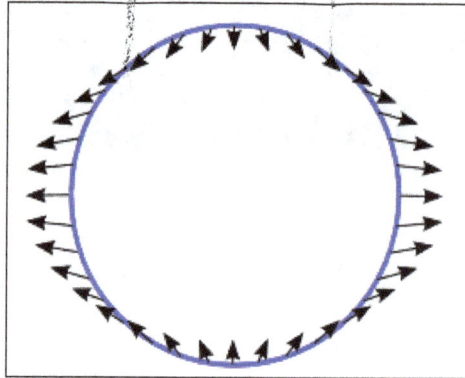

The Moon's (or Sun's) gravity differential field at the surface of the earth is known as the tide generating force. This is the primary mechanism that drives tidal action and explains two tidal equipotential bulges, accounting for two high tides per day.

The ocean's surface is closely approximated by an equipotential surface, (ignoring ocean currents) which is commonly referred to as the geoid. Since the gravitational force is equal to the gradient of the potential, there are no tangential forces on such a surface, and the ocean surface is thus in gravitational equilibrium. Now consider the effect of external, massive bodies such as the Moon and Sun. These bodies have strong gravitational fields that diminish with distance in space and which act to alter the shape of an equipotential surface on the Earth. Gravitational forces follow an inverse-square law (force is inversely proportional to the square of the distance), but tidal forces are inversely proportional to the cube of the distance. The ocean surface moves to adjust to changing tidal equipotential, tending to rise when the tidal potential is high, the part of the Earth nearest the Moon, and the farthest part. When the tidal equipotential changes, the ocean surface is no longer aligned with it, so that the apparent direction of the vertical shifts. The surface then experiences a down slope, in the direction that the equipotential has risen.

Laplace Tidal Equation

The depth of the oceans is much smaller than their horizontal extent; thus, the response to tidal forcing can be modelled using the Laplace tidal equations which incorporate the following features: (1) the vertical (or radial) velocity is negligible, and there is no vertical shear—this is a sheet flow. (2) the forcing is only horizontal (tangential). (3) the Coriolis effect appears as a fictitious lateral forcing proportional to velocity. (4) the rate of change of the surface height is proportional to the negative divergence of velocity multiplied by the depth. The last means that as the horizontal velocity stretches or compresses the ocean as a sheet, the volume thins or thickens, respectively. The boundary conditions dictate no flow across the coastline, and free slip at the bottom. The Coriolis effect steers waves to the right in the northern hemisphere and to the left in the southern allowing coastally trapped waves. Finally, a dissipation term can be added which is an analog to viscosity.

Tidal Amplitude and Cycle Time

The theoretical amplitude of oceanic tides due to the Moon is about 54 cm at the highest point, which corresponds to the amplitude that would be reached if the ocean possessed a uniform depth, there were no landmasses, and the Earth were not rotating. The Sun similarly causes tides, of which the theoretical amplitude is about 25 cm (46% of that of the Moon) with a cycle time of 12 hours. At spring tide the two effects add to each other to a theoretical level of 79 cm, while at neap tide the theoretical level is reduced to 29 cm. Since the orbits of the Earth about the Sun, and the Moon about the Earth, are elliptical, the amplitudes of the tides change somewhat as a result of the varying Earth-Sun and Earth-Moon distances. This causes a variation in the tidal force and theoretical amplitude of about ±18 percent for the Moon and ±5 percent for the Sun. If both the Sun and Moon were at their closest positions and aligned at new moon, the theoretical amplitude would reach 93 cm.

Real amplitudes differ considerably, not only because of variations in ocean depth, and the obstacles to flow caused by the continents, but also because the natural period of wave propagation is of the same order of magnitude as the rotation period: about 30 hours. If there were no land masses, it would take about 30 hours for a long wavelength ocean surface wave to propagate along the equator halfway around the Earth (by comparison, the natural period of the Earth's lithosphere is about 57 minutes).

Tidal Dissipation

The tidal forcing is essentially driven by orbital energy of the Earth Moon system at a rate of about 3.75 Terawatts. The dissipation arises as the basin scale tidal flow drives smaller scale flows which experience turbulent dissipation. This tidal drag gives rise to a torque on the Moon that results in the gradual transfer of angular momentum to its orbit, and a gradual increase in the Earth-Moon separation. As a result of the principle of conservation of angular momentum, the rotational velocity of the Earth is correspondingly slowed. Thus, over geologic time, the Moon recedes from the Earth, at about 3.8 cm/year, and the length of the terrestrial day increases, meaning that there is about 1 less day per 100 million years.

Tidal Observation and Prediction

From ancient times, tides have been observed and discussed with increasing sophistication, first noting the daily recurrence, then its relationship to the Sun and Moon. Eventually the first tide table in China was recorded in 1056 C.E. primarily for the benefit of visitors to see the famous tidal bore in the Qiantang River. In Europe the first known tide-table is thought to be that of John, Abbott of Wallingford, based on high water occurring 48 minutes later each day, and three hours later at London than at the mouth of the Thames. William Thomson led the first systematic harmonic analysis to tidal records starting in 1867. The main result was the building of a tide-predicting machine (TPM) on using a system of pulleys to add together six harmonic functions of time. It was "programmed" by resetting gears and chains to adjust phasing and amplitudes. Similar machines were used until the 1960s.

The first known sea-level record of an entire spring–neap cycle was made in 1831 on the Navy Dock in the Thames Estuary, and many large ports had automatic tide gages stations by 1850. William Whewell first mapped co-tidal lines ending with a nearly global chart in 1836. In order to make these maps consistent, he hypothesized the existence of amphidromes where co-tidal lines meet in the mid-ocean. These points of no tide were confirmed by measurement in 1840 by Captain Hewett, RN, from careful soundings in the North Sea.

Timing

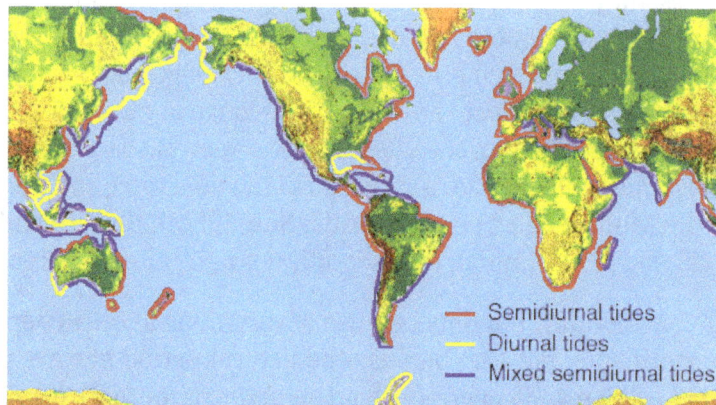

The same tidal forcing has different results depending on
many factors, including coast orientation, continental shelf
margin, water body dimensions.

In most places there is a delay between the phases of the Moon and the effect on the tide. Springs and neaps in the North Sea, for example, are two days behind the new/full Moon and first/third quarter. This is called the age of the tide.

The exact time and height of the tide at a particular coastal point is also greatly influenced by the local bathymetry. There are some extreme cases: the Bay of Fundy, on the east coast of Canada, features the largest well-documented tidal ranges in the world, 16 metres (53 ft), because of the shape of the bay . Southampton in the United Kingdom has a double high tide caused by the interaction between the different tidal harmonics within the region. This is contrary to the popular belief that the flow of water around the Isle of Wight creates two high waters. The Isle of Wight is important, however, as it is responsible for the 'Young Flood Stand', which describes the pause of the incoming tide about three hours after low water.

There are only very slight tides in the Mediterranean Sea and the Baltic Sea owing to their narrow connections with the Atlantic Ocean. Extremely small tides also occur for the same reason in the

Gulf of Mexico and Sea of Japan. On the southern coast of Australia, because the coast is extremely straight (partly due to the tiny quantities of runoff flowing from rivers), tidal ranges are equally small.

Tidal Analysis

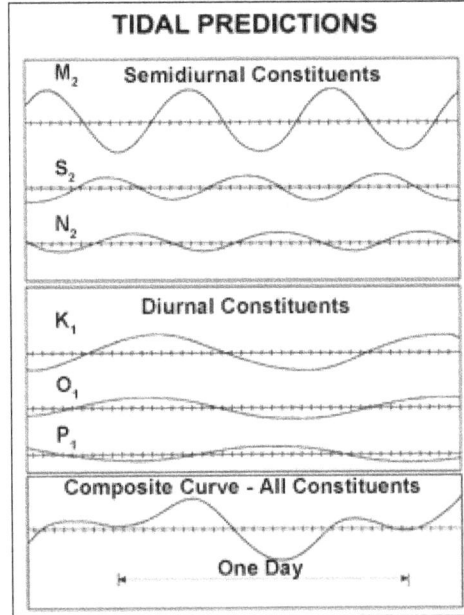

Tidal prediction summing constituent.

Careful Fourier and data analysis over a 19 year period (the National Tidal Datum Epoch in the US) uses carefully selected frequencies called the tidal harmonic constituents. This analysis can be done using only the knowledge of the period of forcing, but without detailed understanding of the physical mathematics, which means that useful tidal tables have been constructed for centuries. The resulting amplitudes and phases can then be used to predict the expected tides. These are usually dominated by the constituents near 12 hours (the semidiurnal constituents), but there are major constituents near 24 hours (diurnal) as well. Longer term constituents are 14 day or fortnightly, monthly, and semiannual. Most coastline is dominated by semidiurnal tides, but some areas such as the South China Sea and the Gulf of Mexico are primarily diurnal. In the semidiurnal areas, the primary constituents M_2(lunar) and S_2(solar) periods differ slightly so that the relative phases, and thus the amplitude of the combined tide, change fortnightly (14 day period).

In the M_2 plot above each cotidal line differs by 1 hour from its neighbors, and the thicker lines show tides in phase with equilibrium at Greenwich. The lines rotate around the amphidromic points counterclockwise in the northern hemisphere so that from Baja California to Alaska and from France to Ireland the M_2 tide propagates northward. In the southern hemisphere this direction is clockwise. On the other hand M_2 tide propagates counterclockwise around New Zealand, but this because the islands act as dam and permit the tides to have different heights on opposite sides of the islands. But the tides do propagate northward on the eastside and southward on the west coast, as predicted by theory. The exception is the Cook Strait where the tidal currents periodically link high to low tide. This is because cotidal lines 180° around the amphidromes are in opposite phase, for example high tide across from low tide. Each tidal constituent has a different pattern of amplitudes, phases, and amphidromic points, so the M_2 patterns cannot be used for other tides.

Tides and Navigation

Tidal flows are of profound importance in navigation and very significant errors in position will occur if they are not taken into account. Tidal heights are also very important; for example many rivers and harbors have a shallow "bar" at the entrance which will prevent boats with significant draft from entering at certain states of the tide.

The timings and velocities of tidal flow can be found by looking at a tidal chart or tidal stream atlas for the area of interest. Tidal charts come in sets, with each diagram of the set covering a single hour between one high tide and another (they ignore the extra 24 minutes) and give the average tidal flow for that one hour. An arrow on the tidal chart indicates the direction and the average flow speed (usually in knots) for spring and neap tides. If a tidal chart is not available, most nautical charts have "tidal diamonds" which relate specific points on the chart to a table of data giving direction and speed of tidal flow.

Standard procedure to counteract the effects of tides on navigation is to: (1) calculate a "dead reckoning" position (or DR) from distance and direction of travel, (2) mark this on the chart (with a vertical cross like a plus sign) and (3) draw a line from the DR in the direction of the tide. The distance the tide will have moved the boat along this line is computed by the tidal speed, and this gives an "estimated position" or EP (traditionally marked with a dot in a triangle).

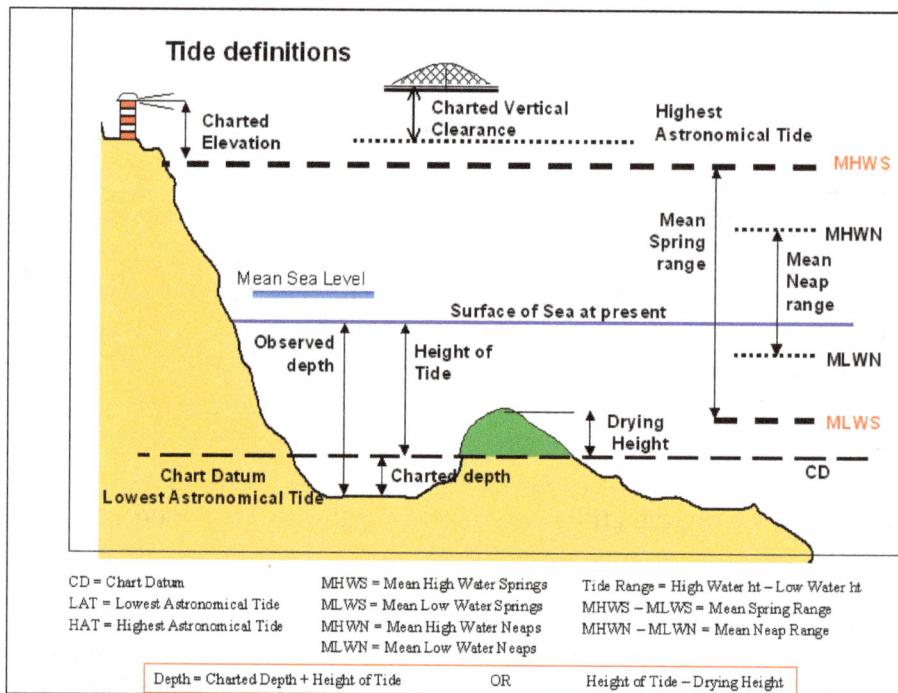

Civil and maritime uses of tidal data.

Nautical charts display the "charted depth" of the water at specific locations with "soundings" and the use of bathymetric contour lines to depict the shape of the submerged surface. These depths are relative to a "chart datum," which is typically the level of water at the lowest possible astronomical tide (tides may be lower or higher for meteorological reasons) and are therefore the minimum water depth possible during the tidal cycle. "Drying heights" may also be shown on the chart, which are the heights of the exposed seabed at the lowest astronomical tide.

Heights and times of low and high tide on each day are published in tide tables. The actual depth of water at the given points at high or low water can easily be calculated by adding the charted depth to the published height of the tide. The water depth for times other than high or low water can be derived from tidal curves published for major ports. If an accurate curve is not available, the rule of twelfths can be used. This approximation works on the basis that the increase in depth in the six hours between low and high tide will follow this simple rule: first hour - 1/12, second - 2/12, third - 3/12, fourth - 3/12, fifth - 2/12, sixth - 1/12.

Biological Aspects

Intertidal Ecology

A rock, seen at low tide, exhibiting
typical intertidal zonation.

Intertidal ecology is the study of intertidal ecosystems, where organisms live between the low and high tide lines. At low tide, the intertidal is exposed (or 'emersed') whereas at high tide, the intertidal is underwater (or 'immersed'). Intertidal ecologists therefore study the interactions between intertidal organisms and their environment, as well as between different species of intertidal organisms within a particular intertidal community. The most important environmental and species interactions may vary based on the type of intertidal community being studied, the broadest of classifications being based on substrates - rocky shore and soft bottom communities.

Organisms living in this zone have a highly variable and often hostile environment, and have evolved various adaptations to cope with, and even exploit, these conditions. One easily visible feature of intertidal communities is vertical zonation, where the community is divided into distinct vertical bands of specific species going up the shore. Species ability to cope with desiccation determines their upper limits, while competition with other species sets their lower limits.

Intertidal regions are utilized by humans for food and recreation, but anthropogenic actions also have major impacts, with overexploitation, invasive species and climate change being among the problems faced by intertidal communities. In some places Marine Protected Areas have been established to protect these areas and aid in scientific research.

Biological Rhythyms and the Tides

Intertidal organisms are greatly affected by the approximately fortnightly cycle of the tides, and hence their biological rhythms tend to occur in rough multiples of this period. This is seen not only in the intertidal organisms however, but also in many other terrestrial animals, such as the vertebrates. Examples include gestation and the hatching of eggs. In humans, for example, the menstrual cycle lasts roughly a month, an even multiple of the period of the tidal cycle. This may be evidence of the common descent of all animals from a marine ancestor.

Other Tides

In addition to oceanic tides, there are atmospheric tides as well as earth tides. All of these are continuum mechanical phenomena, the first two being fluids and the third solid (with various modifications).

Atmospheric tides are negligible from ground level and aviation altitudes, drowned by the much more important effects of weather. Atmospheric tides are both gravitational and thermal in origin, and are the dominant dynamics from about 80 km to 120 km where the molecular density becomes too small to behave as a fluid.

Earth tides or terrestrial tides that affect the entire rocky mass of the Earth. The Earth's crust shifts (up/down, east/west, north/south) in response to the Moon's and Sun's gravitation, ocean tides, and atmospheric loading. While negligible for most human activities, the semidiurnal amplitude of terrestrial tides can reach about 55 cm at the equator (15 cm is due to the Sun) which is important in GPS calibration and VLBI measurements. Also to make precise astronomical angular measurements requires knowledge of the earth's rate of rotation and nutation, both of which are influenced by earth tides. The semi-diurnal M_2 Earth tides are nearly in phase with the Moon with tidal lag of about two hours.

Terrestrial tides also need to be taken in account in the case of some particle physics experiments. For instance, at the CERN or SLAC, the very large particle accelerators were designed while taking terrestrial tides into account for proper operation. Among the effects that need to be taken into account are for circular accelerators and particle beam energy Since tidal forces generate currents of conducting fluids within the interior of the Earth, they affect in turn the Earth's magnetic field itself.

The *galactic tide* is the tidal force exerted by galaxies on stars within them and satellite galaxies orbiting them. The effects of the galactic tide on the Solar System's Oort cloud are believed to be the cause of 90 percent of all observed long-period comets.

When oscillating tidal currents in the stratified ocean flow over uneven bottom topography, they generate internal waves with tidal frequencies. Such waves are called internal tides.

Misnomers

Tsunamis, the large waves that occur after earthquakes, are sometimes called tidal waves, but this name is due to their resemblance to the tide, rather than any actual link to the tide itself. Other phenomena unrelated to tides but using the word tide are rip tide, storm tide, hurricane tide, and black tide, referring to oil spills; or red tides, that refer to algae blooms.

Ocean Waves

Ocean wave is a disturbance of the ocean's surface. Because of their great mobility, water particles easily come out of equilibrium and oscillate under the influence of various kinds of forces. Waves are caused by the tide-forming forces of the moon and sun, by winds, by fluctuations in atmospheric pressure, by underwater earthquakes, and by deformations of the ocean floor.

According to their cause, sea waves are subdivided into normal tide waves, waves caused by wind, waves caused by fluctuations in atmospheric pressure (seiches), and seismic waves (tsunamis). In most cases, wave motions are characterized by irregular shape. It is necessary to distinguish between the shifting of particles within the wave and the apparent motion of the wave's shape, which consists in the movement of its profile through space. The particles involved in the wave are moving along closed or almost closed trajectories.

The main characteristics of sea waves are height, which is the vertical distance between the crest and the trough of the wave; length, which is the horizontal distance between two adjacent crests or troughs; the velocity of motion of the wave shape, or phase velocity; and the period. The period of waves caused by wind does not exceed 30 sec; those caused by atmospheric-pressure variations or by earthquakes have periods measured in minutes, dozens of minutes, or hours. The periods of normal tide waves are expressed in hours.

Depending on the prevailing role of the forces acting in the formation of wave motions, waves are subdivided into gravitational and capillary types. Waves that continue to exist after the end of the action of the forces that caused them are called free waves, as distinct from induced waves, which are maintained by an uninterrupted energy input.

Waves caused by wind are of the induced type. They are formed by the energy of the wind through direct pressure of the airflow on the windward sides of the crests and through its friction on the wave's surface. The development of waves caused by wind begins with the formation of ripples, which are capillary waves. As the capillary waves grow, they change into gravitational waves, which gradually increase in length and height. In their initial stage of development, the waves run in parallel lines, which later disintegrate into individual crests (three-dimensional wave pattern). The surface of the water stirred up by the wind takes on a very complicated shape, which continually changes over time. Waves caused by wind are always present on the surface of the sea, and their dimensions are most varied (sometimes up to 400 m long and 12-13 m high, with speeds of 14-15 m/sec).

In a deep sea the size and pattern of the waves are determined by the wind velocity, the duration of its action, the"wave race" (the distance from the leeward shore to the observation point along the wind direction), and the structure of the wind field and the configuration of the shoreline. In a shallow sea, the depth of the sea and the relief of its bottom also have an effect on the process of wave formation. Shallow seas tend to limit the growth of waves. If the wind that has caused the waves abates, the waves caused by it gradu-ally change into free waves, called swells, whose waves have a more regular shape than those caused by wind and have a greater length between crests. The most frequent case is a mixed wave pattern in which ripples and waves caused by wind occur simultaneously.

The study of sea waves is of great practical interest in connection with numerous problems in navigation, marine hydraulic-engineering construction, and shipbuilding. It requires detailed theoretical and experimental research, using various instruments installed on vessels and on shore.

Among the devices used for observing sea waves are wave-measuring rods or marks, wavemeters, and wave graphs of various systems. Stereophotography makes possible the recording of surface conditions over wide areas. Long-period sea waves—for example, tide waves—are recorded by devices called tide gauges.

Types of Ocean Waves and Wave Classification

Ocean waves can be classified in several ways. The most intuitive and commonly used classification is based on the wave period or the associated wavelength. In Table, a summary of the different types of surface waves is presented with respect to wave period. The associated originating forcing and restoring mechanisms are also reported. A graphical representation is provided in Figure, where an idealized wave energy spectrum shows the full range of ocean wave components.

Table: Ocean Wave Classification.

Classification	Period band	Generating forces	Restoring forces
Capillary waves	<0.1 s	Wind	Surface tension
Ultragravity waves	0.1–1 s	Wind	Surface tension and gravity
Gravity waves	1–20 s	Wind	Gravity
Infragravity waves	20 s to 5 min	Wind and atmospheric pressure gradients	Gravity
Long-period waves	5 min to 12 h	Atmospheric pressure gradients and earthquake	Gravity
Ordinary tidal waves	12–24 h	Gravitational attraction	Gravity and Coriolis force
Transtidal waves	>24 h	Storms and gravitational attraction	Gravity and Coriolis force

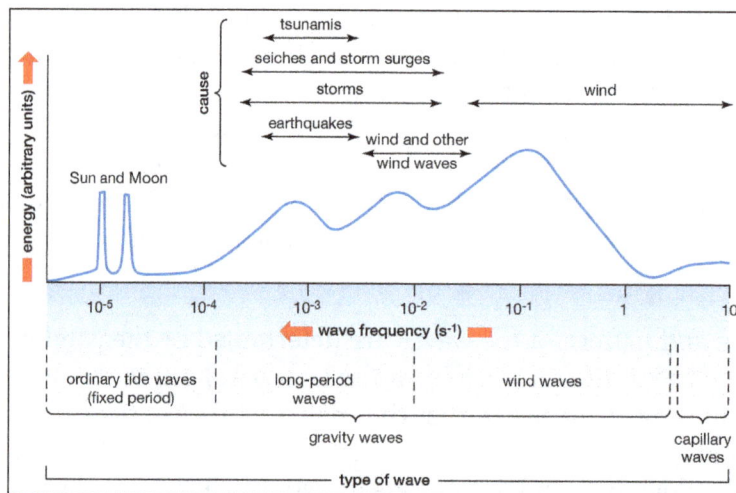

Frequency and period of ocean waves.

Capillary Waves

The shortest-period waves, and the first to be noticed on the ocean surface when wind starts blowing, are the capillary waves, which resemble cat's paws ripping the otherwise smooth surface. This peculiar wavy structure is generally forced by a light breeze of speeds of about 3 m/s (taken at a reference height of 10 m from the water level) and assumes a fine structure of small ripples with a wavelength of less than 1.5 cm and period less than 0.1 s.

Example of capillary waves.

The dynamics of capillary waves is dominated primarily by surface tension, which forces group velocity (the speed at which energy propagates) to be 1.5 times greater than the phase velocity. As waves keep growing under the influence of wind, however, the initially small ripples evolve into longer waves. For wavelength of approximately 1.7 cm (or wave period of about 0.33 s), gravity cancels capillary effects, suppressing dispersion. At this stage of wave growth, wave groups and wave phases propagate at the same speed. Immediately above this threshold, gravity effects dominate wave dynamics, while surface tension only plays a secondary role. The resulting oscillations are normally classified as an ultragravity wave.

Gravity Waves: Wind Sea and Swell

A consistent blowing of wind over a substantial fetch (i.e., the distance over which the wind blows) forces waves to become much longer than the threshold wavelength of 1.7 cm. As the wavelength grows longer than 1.5 m (i.e., wave period becomes larger than 1 s), surface tension becomes negligible and gravity remains the sole restoring mechanism. Under these circumstances, waves are classified as gravity waves. It is worth mentioning, in this regard, that gravity acts on wave dispersion by inducing wave phases to propagate faster than wave groups and thus reversing the effect of surface tension. Generally speaking, gravity waves assume periods ranging from a minimum of about 1 s up to maximum of approximately 25 s (i.e., wavelength varies roughly between 1.5 and 900 m).

Under the direct effect of the local wind, a large number of components with different wave periods, direction of propagations and phases are generated. The resulting wave field is an interaction of all these components, which generates an erratic (irregular) pattern normally known as wind sea. Despite the wide period band of wind-generated waves, the dominant components of wind sea remain relatively short. During severe storm conditions, for example, the wave period increases up to maximum of approximately 10–12 s (or associated wavelengths of 150–220 m). The wave height, however, grows substantially, steepening the wave profile. This increases wave-induced velocities on and below the water surface, enhancing loads on marine structures and contributing to the mixing of the upper ocean by injecting turbulence directly throughout a depth comparable with the wavelength. If waves become sufficiently steep and wave components propagate over a narrow range of directions, the instability of wave groups to side-band perturbations can cause a rapid and substantial growth of wave amplitude at the expense of the surrounding waves, leading to the formation of extremely large waves. When the wave height of such events exceeds twice the significant wave height (i.e., the average of the highest 1/3 of the waves in a 20-min wave record), a freak or rogue wave is normally identified. In this regard, a well-documented example of a rogue wave is the "New Year" wave. This event was recorded at the Draupner oil field (North Sea) on 1 January 1995, and consisted of a 25.6 m high wave in a background sea state of about 12 m significant wave height (i.e., the wave height exceeded 2.1 times the significant wave height). Despite their low probability of occurrence, these extraordinary wave events may sometime represent a severe sea hazard for all types of engineering activities, especially during an already severe storm condition.

When waves propagate over a depth that is much larger than the wavelength (i.e., the water depth can be considered as of infinite depth), longer waves travel faster than shorter ones, dispersing from one another. As a consequence, long waves rapidly move outside the generating area and become known as swells. Swells have a typical wavelength that is greater than 260 m (i.e., period larger than 13 s) up to maximum of approximately 900 m. As their height is normally small, dissipation is less intense if compared with wind sea. As an example, a long wave with period larger than 13 s and small amplitude loses about half of its energy over a distance of about 20,000 km. It is not uncommon, therefore, that a swell that generates in the Antarctic Ocean travels all the way to the Alaska with very little energy dissipation. However, like all gravity waves, swell grows in amplitude and decreases in wavelength over rapidly shoaling regions before it breaks nearby the shore, quickly losing energy.

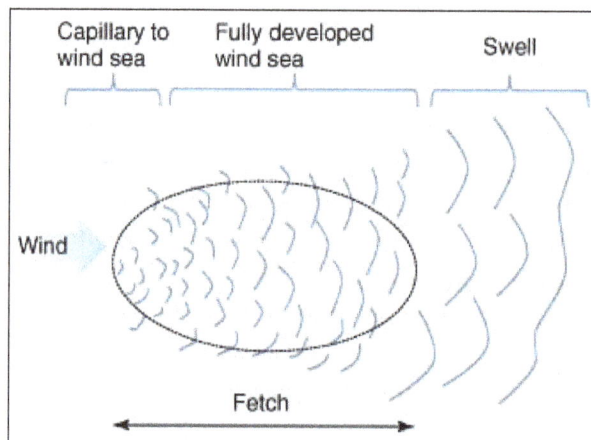

Schematic representation of wind sea and swell.

Infragravity Waves

Nonlinear interactions between wave components convert part of the energy associated to wind-generated gravity waves into subharmonics with periods ranging from about 20 to 30 s up to a maximum of approximately 5 min. These long oscillations, which are driven primarily by swell, are bound to the generating wave trains and are normally known as infragravity waves. An example of measured infragravity waves is presented in Figure, where the infragravity component was extracted from an original time series (solid line in the figure) by filtering period lower than 30 s.

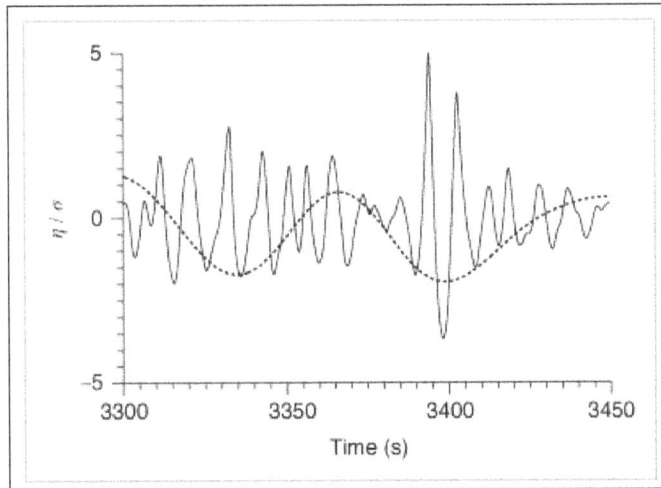

Example of bound infragravity waves: recorded time series
(solid line) and filtered infragravity waves (dashed line).

In coastal areas, incident wind sea and swells dissipate their energy in the form of depth-induced breaking, releasing the less-energetic infragravity components. Note that as the waves are no longer bounded to the originating wave packets, they are thus defined as free waves. Due to their small amplitude, infragravity waves do not reach the breaking limit and hence propagate toward the shore until they are reflected. The resulting seaward-propagating free infragravity waves may be further reflected backward from a turning point on the sloping beach or radiate into the deep ocean.

Generally, infragravity motion observed on the continental shelf in depths ranging from 8 to 200 m is a mixture of subharmonics bounded to the incident waves and free waves radiated from the shore, which are normally more energetic than the former. Infragravity waves may affect substantially sediment transport and other coastal processes and activities, including port operations and moorings due to induced harbor oscillations.

Long-period Waves (Tsunamis, Seiches and Storm Surges)

Well-defined waves with periods longer than 5 min are routinely recorded in the ocean (Nielsen, 2009). Although different originating mechanisms can be responsible for such waves, meteorological conditions and earthquakes remain the primary cause. Normally, long oscillations generated by atmospheric conditions are known as seiches and storm surges, while tsunamis identify waves originated from earthquakes. Despite the long wavelength, the restoring mechanism is still dominated by gravity.

Tsunamis are long waves with period varying between 1 and 20 min (wavelength from a few kilometers up to a few hundreds of kilometers) that are generated by sudden tectonic changes to the sea bed or landslides that are usually attributed to earthquakes and submarine volcanic activity. In the open ocean, tsunamis have very small amplitude (only rarely wave height exceeds 1 m) and generally pass completely unnoticed. Propagation into shallower waters, however, makes wave shoal, compressing the shape of the oscillation. As a result, its speed diminishes of about one order of magnitude (from about 800 to <80 km/h), while its wavelength reduces to less than 20 km with a consequent substantial growth of wave height. Except for the largest tsunamis, the approaching wave does not break, but rather appears like a fast-moving tidal bore. It is interesting to note that tsunamis may also feature multiple waves with significant time (normally hours) between arrivals of subsequent wave crests. The first wave reaching the shore may not necessarily have the highest run up, though.

It is not uncommon for seismic activities and tsunamis to generate additional long-waves components (seiches) due to local geographical peculiarities. The primary causes of seiches, nonetheless, remains related to meteorological disturbances such as wind gusts or atmospheric pressure variations, which induce a resonance effect on enclosed or partially enclosed water basins (e.g., closed seas such as the Adriatic Sea, the Baltic Sea, and the North Sea) and originate very long oscillation.

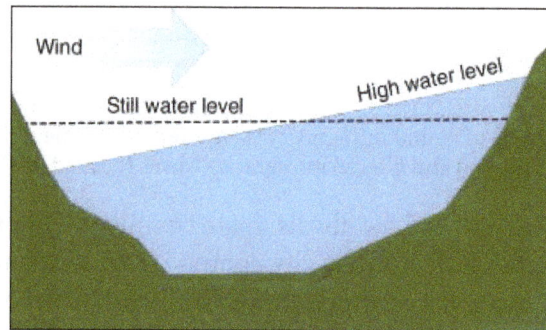

A schematic representation of seiches in a closed water basin.

The range of natural periods for seiches is rather large and strongly depends on the fundamental resonance period of the water basin. In most of the seiches that occur in nature, the period T can be estimated as $T = 2\,L/(gh)^{0.5}$, where g is the acceleration due to gravity, L the length of the basin, and h the average depth of the basin. Normally, seiches induced by wind gusts have wavelength of a few kilometers with an associated wave period of the order of undress seconds. Variations in atmospheric pressure or mean wind direction due to moving weather systems, on the other hand, are responsible for oscillation of much larger scales with associated period of the order of hours.

Large atmospheric low-pressure systems (e.g., tropical cyclones or extra-tropical storms) cause much longer oscillations, namely storm surges, by raising the water level in regions of low pressure and dropping it in regions of high pressure. For a stationary system, for example, an increase of sea level of about 1 cm corresponds to a variation of atmospheric pressure of approximately 1 hPa. The wave period of storm surges is normally of the same order of magnitude of the meteorological disturbance, that is, it varies from a few hours to a few days.

Due to the large wavelength, long waves and especially storm surges are normally perceived as a change in the water level rather than a surface water wave, especially near shore. In the most extreme cases, this can cause flooding of coastal areas and disruptions to all coastal activities.

Equatorial Wave

Equatorial waves are oceanic and atmospheric waves trapped close to the equator, meaning that they decay rapidly away from the equator, but can propagate in the longitudinal and vertical directions. Wave trapping is the result of the Earth's rotation and its spherical shape which combine to cause the magnitude of the Coriolis force to increase rapidly away from the equator. Equatorial waves are present in both the tropical atmosphere and ocean and play an important role in the evolution of many climate phenomena such as El Niño. Many physical processes may excite equatorial waves including, in the case of the atmosphere, diabatic heat release associated with cloud formation, and in the case of the ocean, anomalous changes in the strength or direction of the trade winds.

Equatorial waves may be separated into a series of subclasses depending on their fundamental dynamics (which also influences their typical periods and speeds and directions of propagation). At shortest periods are the equatorial gravity waves while the longest periods are associated with the equatorial Rossby waves. In addition to these two extreme subclasses, there are two special subclasses of equatorial waves known as the mixed Rossby-gravity wave (also known as the Yanai wave) and the equatorial Kelvin wave. The latter two share the characteristics that they can have any period and also that they may carry energy only in an eastward (never westward) direction.

Equatorial Rossby and Rossby-gravity Waves

Rossby-gravity waves, first observed in the stratosphere by M. Yanai, always carry energy eastward. But, oddly, their 'crests' and 'troughs' may propagate westward if their periods are long enough. The eastward speed of propagation of these waves can be derived for an inviscid slowly moving layer of fluid of uniform depth H. Because the Coriolis parameter ($f = 2\Omega \sin(\theta)$ where Ω is the angular velocity of the earth, 7.2921×10^{-5} rad/s, and θ is latitude) vanishes at 0 degrees latitude (equator), the "equatorial beta plane" approximation must be made. This approximation states that "f" is approximately equal to βy, where "y" is the distance from the equator and "β" is the variation of the coriolis parameter with latitude $\dfrac{\partial f}{\partial y} = \beta$. With the inclusion of this approximation, the governing equations become (neglecting friction):

- The continuity equation (accounting for the effects of horizontal convergence and divergence and written with geopotential height):

$$\frac{\partial \phi}{\partial t} + c^2 \left(\frac{\partial v}{\partial y} + \frac{\partial u}{\partial x} \right) = 0$$

- The u-momentum equation (zonal wind component):

$$\frac{\partial u}{\partial t} - v\beta y = -\frac{\partial \phi}{\partial x}$$

- The v-momentum equation (meridional wind component):

$$\frac{\partial v}{\partial t} + u\beta y = -\frac{\partial \phi}{\partial y}.$$

We may seek travelling-wave solutions of the form:

$$\{u, v, \phi\} = \{\hat{u}(y), \hat{v}(y), \hat{\phi}(y)\} e^{i(kx - \omega t)}.$$

Substituting this exponential form into the three equations above, and eliminating u and ϕ leaves us with an eigenvalue equation:

$$-\frac{\partial^2 \hat{v}}{\partial y^2} + \left(\frac{\beta^2}{c^2}\right) y^2 \, \hat{v} = \left(\frac{\omega^2}{c^2} - k^2 - \frac{\beta k}{\omega}\right) \hat{v}.$$

for $\hat{v}(y)$. Recognizing this as the Schrödinger equation for a quantum harmonic oscillator of frequency $\Omega = \beta / c$, we know that we must have:

$$\left(\frac{\omega^2}{c^2} - k^2 - \frac{\beta k}{\omega}\right) = \frac{\beta}{c}(2n+1), \quad n \geq 0$$

For the solutions to tend to zero away from the equator. For each integer n therefore, this last equation provides a dispersion relation linking the wavenumber k to the angular frequency ω.

In the special case n = 0 the dispersion equation reduces to,

$$(\omega + ck)(\omega^2 - ck\omega - c\beta) = 0,$$

but the root $\omega = -ck$ has to be discarded because we had to divide by this factor in eliminating u, ϕ. The remaining pair of roots correspond to the *Yanai* or mixed Rossby-gravity mode whose group velocity is always to the east and interpolates between two types of $n > 0$ modes: the higher frequency Poincaré gravity waves whose group velocity can be to the east or to the west, and the low-frequency equatorial Rossby waves whose dispersion relation can be approximated as,

$$\omega = \frac{-\beta k}{k^2 + \beta(2n+1)/c}.$$

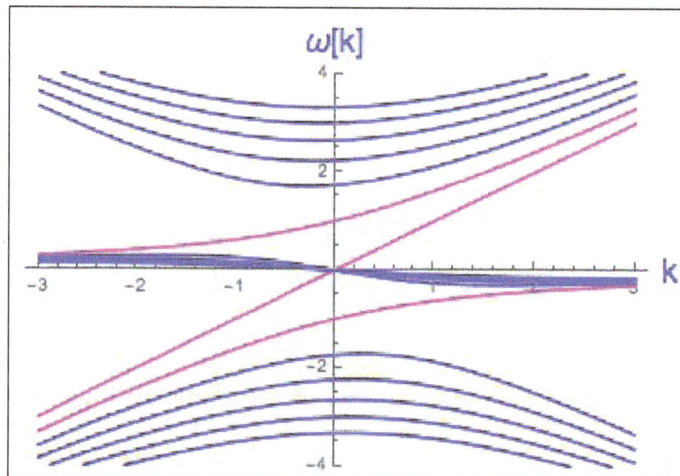

Dispersion relations for equatorial waves with different values of n: The dense narrow band of low-frequency Rossby waves and the higher frequency Poincaré gravity waves are in blue. The topologically protected Kelvin and Yanai modes are highlighted in magenta.

The Yanai modes, together with the Kelvin waves are rather special in that they are *topologically protected*. Their existence is guaranteed by the fact that the band of positive frequency Poincaré modes in the f-plane form a non-trivial bundle over the two-sphere $\sqrt{k^2 + f^2} = 1$. This bundle is characterized by Chern number $c_1 = 2$. The Rossby waves have $c_1 = 0$ and the negative frequency Poincaré modes have $c_1 = -2$. Through the bulk-boundary connection this necessitates the existence of two modes (Kelvin and Yanai) that cross the frequency gaps between the Poincaré and Rossby bands and are localized near the equator where $f = \beta y$ changes sign.

Equatorial Kelvin Waves

Discovered by Lord Kelvin, coastal Kelvin waves are trapped close to coasts and propagate along coasts in the Northern Hemisphere such that the coast is to the right of the alongshore direction of propagation (and to the left in the Southern Hemisphere). Equatorial Kelvin waves behave somewhat as if there were a wall at the equator – so that the equator is to the right of the direction of along-equator propagation in the Northern Hemisphere and to the left of the direction of propagation in the Southern Hemisphere, both of which are consistent with eastward propagation along the equator. The governing equations for these equatorial waves are similar to those presented above, except that there is no meridional velocity component v (y) (that is, no flow in the north–south direction).

- The continuity equation (accounting for the effects of horizontal convergence and divergence):

$$\frac{\partial \phi}{\partial t} + c^2 \frac{\partial u}{\partial x} = 0$$

- The u-momentum equation (zonal wind component):

$$\frac{\partial u}{\partial t} = -\frac{\partial \phi}{\partial x}$$

- The v-momentum equation (meridional wind component):

$$u\beta y = -\frac{\partial \phi}{\partial y}$$

The solution to these equations yields the following phase speed: $c^2 = gH$; this result is the same speed as for shallow-water gravity waves without the effect of Earth's rotation. Therefore, these waves are non-dispersive (because the phase speed is not a function of the zonal wavenumber). Also, these Kelvin waves only propagate towards the east (because as Φ approaches zero, y approaches infinity).

Connection to El Niño Southern Oscillation

Kelvin waves have been connected to El Niño (beginning in the Northern Hemisphere winter months) in recent years in terms of precursors to this atmospheric and oceanic phenomenon. Many scientists have utilized coupled atmosphere–ocean models to simulate an El Niño Southern Oscillation (ENSO) event and have stated that the Madden–Julian oscillation (MJO) can trigger oceanic Kelvin waves throughout its 30- to 60-day cycle or the latent heat of condensation can be released (from intense convection) resulting in Kelvin waves as well; this process can then signal the onset of an El Niño event. The weak low pressure in the Indian Ocean (due to the MJO) typically propagates eastward into the North Pacific Ocean and can produce easterly winds. These easterly winds can transfer West Pacific warm water toward the east, thereby exciting a Kelvin wave, which in this sense can be thought of as a warm-water anomaly that travels under the ocean's surface at a depth of about 150 meters. This wave can be observed at the surface by a slight rise in sea surface height of about 8 cm (associated with a depression of the thermocline) and an SST increase that covers hundreds of square kilometres across the surface of the ocean.

If the Kelvin wave hits the South American coast (specifically Ecuador), its warm water gets transferred upward, which creates a large warm pool at the surface. That warm water also starts to flow southward along the coast of Peru and north towards Central America and Mexico, and may reach parts of Northern California; the wave can then be tracked primarily using an array of 70 buoys anchored along the entire width of equatorial Pacific Ocean, from Papua New Guinea to the Ecuador coast. Temperature sensors are placed at different depths along the buoys' anchor-lines and are then able to record sub-surface water temperature. The sensors send their data in real-time via satellite to a central processing facility. These temperature measurements are then compared and contrasted to historically and seasonally adjusted average water temperatures for each buoy location. Some results indicate deviations from the 'normal' expected temperatures. Such deviations are referred to as anomalies and can be thought of as either warmer-than-normal (El Niño) or cooler-than-normal (La Niña) conditions.

The overall ENSO cycle can be explained as follows (in terms of the wave propagation throughout the Pacific Ocean): ENSO begins with a warm pool traveling from the western Pacific to the eastern Pacific in the form of Kelvin waves (the waves carry the warm SSTs) that resulted from the MJO. After approximately 3 to 4 months of propagation across the Pacific (along the equatorial region), the Kelvin waves reach the western coast of South America and interact (merge/mix) with the cooler Peru current system. This causes a rise in sea levels and sea level temperatures in the general region. Upon reaching the coast, the water turns to the north and south and results in El Niño conditions to the south. Because of the changes in sea-level and sea-temperature due to the Kelvin waves, an infinite number of Rossby waves are generated and move back over the Pacific. Rossby waves then enter the equation and, as previously stated, move at lower velocities than the Kelvin waves and can take anywhere from nine months to four years to fully cross the Pacific Ocean basin (from boundary to boundary). And because these waves are equatorial in nature, they decay rapidly as distance from the equator increases; thus, as they move away from the equator, their speed decreases as well, resulting in a wave delay. When the Rossby waves reach the western Pacific they ricochet off the coast and become Kelvin waves and then propagate back across the Pacific in the direction of the South America coast. Upon return, however, the waves decrease the sea-level (reducing the depression in the thermocline) and sea surface temperature, thereby returning the area to normal or sometimes La Niña conditions.

In terms of climate modeling and upon coupling the atmosphere and the ocean, an ENSO model typically contains the following dynamical equations:

- 3 primitive equations for the atmosphere (as mentioned above) with the inclusion of frictional parameterizations: 1) u-momentum equation, 2) v-momentum equation, and 3) continuity equation.

- 4 primitive equations for the ocean (as stated below) with the inclusion of frictional parameterizations:

  ○ U-momentum:

  $$\frac{\partial u}{\partial t} - v\beta y = \frac{\tau_x}{\rho h}$$

  ○ V-momentum:

  $$\frac{\partial v}{\partial t} - u\beta y = \frac{\tau_y}{\rho h}$$

  ○ Continuity:

  $$\frac{\partial h}{\partial t} + h\left(\frac{\partial u}{\partial x} + \frac{\partial v}{\partial y}\right) - K_E T = 0$$

  ○ Thermodynamic energy:

  $$\frac{\partial T}{\partial t} + u\frac{\partial T}{\partial x} - K_T h = 0$$

Note that h is the depth of the fluid (similar to the equivalent depth and analogous to H in the primitive equations listed above for Rossby-gravity and Kelvin waves), K_T is temperature diffusion, K_E is eddy diffusivity, and τ is the wind stress in either the x or y directions.

Kelvin Wave

A Kelvin wave is a wave in the ocean or atmosphere that balances the Earth's Coriolis force against a topographic boundary such as a coastline, or a waveguide such as the equator. A feature of a Kelvin wave is that it is non-dispersive, i.e., the phase speed of the wave crests is equal to the group speed of the wave energy for all frequencies. This means that it retains its shape as it moves in the alongshore direction over time.

A Kelvin wave (fluid dynamics) is also a long scale perturbation mode of a vortex in superfluid dynamics; in terms of the meteorological or oceanographical derivation, one may assume that the meridional velocity component vanishes (i.e. there is no flow in the north–south direction, thus making the momentum and continuity equations much simpler). This wave is named after the discoverer, Lord Kelvin.

Coastal Kelvin Wave

In a stratified ocean of mean depth H, free waves propagate along coastal boundaries (and hence become trapped in the vicinity of the coast itself) in the form of internal Kelvin waves on a scale of about 30 km. These waves are called coastal Kelvin waves, and have propagation speeds of approximately 2 m/s in the ocean. Using the assumption that the cross-shore velocity v is zero at the coast, $v = 0$, one may solve a frequency relation for the phase speed of coastal Kelvin waves,

which are among the class of waves called boundary waves, edge waves, trapped waves, or surface waves (similar to the Lamb waves). The (linearised) primitive equations then become the
following:

- The continuity equation (accounting for the effects of horizontal convergence and divergence):

$$\frac{\partial u}{\partial x} + \frac{\partial v}{\partial y} = \frac{-1}{H}\frac{\partial \eta}{\partial t}$$

- The u-momentum equation (zonal wind component):

$$\frac{\partial u}{\partial t} = -g\frac{\partial \eta}{\partial x} + fv$$

- The v-momentum equation (meridional wind component):

$$\frac{\partial v}{\partial t} = -g\frac{\partial \eta}{\partial y} - fu.$$

If one assumes that the Coriolis coefficient f is constant along the right boundary conditions and
the zonal wind speed is set equal to zero, then the primitive equations become the following:

- The continuity equation:

$$\frac{\partial v}{\partial y} = \frac{-1}{H}\frac{\partial \eta}{\partial t}$$

- The u-momentum equation:

$$g\frac{\partial \eta}{\partial x} = fv$$

- The v-momentum equation:

$$\frac{\partial v}{\partial t} = -g\frac{\partial \eta}{\partial y}.$$

The solution to these equations yields the following phase speed: $c^2 = gH$, which is the same speed
as for shallow-water gravity waves without the effect of Earth's rotation. It is important to note
that for an observer traveling with the wave, the coastal boundary (maximum amplitude) is always
to the right in the northern hemisphere and to the left in the southern hemisphere (i.e. these waves
move equatorward – negative phase speed – on a western boundary and poleward – positive phase
speed – on an eastern boundary; the waves move cyclonically around an ocean basin).

Equatorial Kelvin wave

The equatorial zone essentially acts as a waveguide, causing disturbances to be trapped in the vicinity of the Equator, and the equatorial Kelvin wave illustrates this fact because the Equator acts

analogously to a topographic boundary for both the Northern and Southern Hemispheres, making this wave very similar to the coastally-trapped Kelvin wave. The primitive equations are identical to those used to develop the coastal Kelvin wave phase speed solution (U-momentum, V-momentum, and continuity equations) and the motion is unidirectional and parallel to the Equator. Because these waves are equatorial, the Coriolis parameter vanishes at 0 degrees; therefore, it is necessary to use the equatorial beta plane approximation that states:

$$f = \beta y,$$

Where β is the variation of the Coriolis parameter with latitude. This equatorial beta plane assumption requires a geostrophic balance between the eastward velocity and the north-south pressure gradient. The phase speed is identical to that of coastal Kelvin waves, indicating that the equatorial Kelvin waves propagate toward the east without dispersion (as if the earth were a non-rotating planet). For the first baroclinic mode in the ocean, a typical phase speed would be about 2.8 m/s, causing an equatorial Kelvin wave to take 2 months to cross the Pacific Ocean between New Guinea and South America; for higher ocean and atmospheric modes, the phase speeds are comparable to fluid flow speeds.

When the motion at the Equator is to the east, any deviation toward the north is brought back toward the Equator because the Coriolis force acts to the right of the direction of motion in the Northern Hemisphere, and any deviation to the south is brought back toward the Equator because the Coriolis force acts to the left of the direction of motion in the Southern Hemisphere. Note that for motion toward the west, the Coriolis force would not restore a northward or southward deviation back toward the Equator; thus, equatorial Kelvin waves are only possible for eastward motion (as noted above). Both atmospheric and oceanic equatorial Kelvin waves play an important role in the dynamics of El Nino-Southern Oscillation, by transmitting changes in conditions in the Western Pacific to the Eastern Pacific.

There have been studies that connect equatorial Kelvin waves to coastal Kelvin waves. Moore (1968) found that as an equatorial Kelvin wave strikes an "eastern boundary", part of the energy is reflected in the form of planetary and gravity waves; and the remainder of the energy is carried poleward along the eastern boundary as coastal Kelvin waves. This process indicates that some energy may be lost from the equatorial region and transported to the poleward region.

Equatorial Kelvin waves are often associated with anomalies in surface wind stress. For example, positive (eastward) anomalies in wind stress in the central Pacific excite positive anomalies in 20 °C isotherm depth which propagate to the east as equatorial Kelvin waves.

Capillary Wave

A capillary wave is a wave traveling along the phase boundary of a fluid, whose dynamics and phase velocity are dominated by the effects of surface tension. Capillary waves are common in nature, and are often referred to as ripples. The wavelength of capillary waves on water is typically less than a few centimeters, with a phase speed in excess of 0.2–0.3 meter/second.

A longer wavelength on a fluid interface will result in gravity–capillary waves which are influenced by both the effects of surface tension and gravity, as well as by fluid inertia. Ordinary gravity waves have a still longer wavelength.

When generated by light wind in open water, a nautical name for them is cat's paw waves. Light breezes which stir up such small ripples are also sometimes referred to as cat's paws. On the open ocean, much larger ocean surface waves (seas and swells) may result from coalescence of smaller wind-caused ripple-waves.

Capillary wave (ripple) in water.

Ripples on Lifjord in Øksnes, Norway.

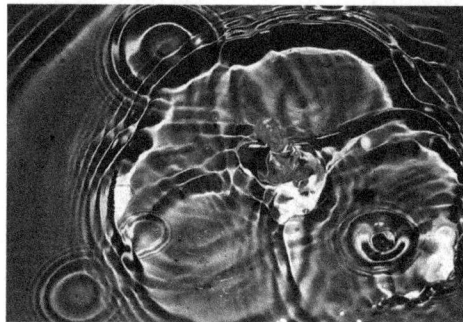

Capillary waves produced by droplet impacts
on the interface between water and air.

Dispersion Relation

The dispersion relation describes the relationship between wavelength and frequency in waves. Distinction can be made between pure capillary waves – fully dominated by the effects of surface tension – and gravity–capillary waves which are also affected by gravity.

Capillary Waves and Proper

The dispersion relation for capillary waves is,

$$\omega^2 = \frac{\sigma}{\rho + \rho'} |k|^3,$$

Where ω is the angular frequency, σ the surface tension, ρ the density of the heavier fluid, ρ' the density of the lighter fluid and k the wavenumber. The wavelength is $\lambda = \frac{2\pi}{k}$. For the boundary between fluid and vacuum (free surface), the dispersion relation reduces to:

$$\omega^2 = \frac{\sigma}{\rho} |k|^3.$$

Gravity–capillary Waves

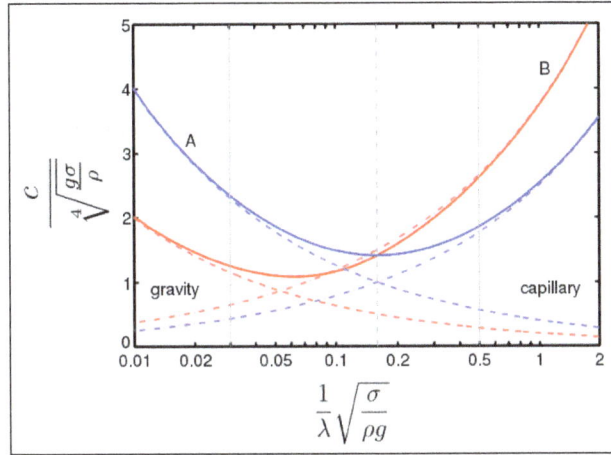

Dispersion of gravity–capillary waves on the surface of deep water (zero mass density of upper layer, $\rho = 0$). Phase and group velocity divided by $\sqrt[4]{g\sigma / \rho}$ as a function of inverse relative wavelength $\frac{1}{\lambda}\sqrt{\sigma / (\rho g)}$.

- Blue lines (A): phase velocity, Red lines (B): group velocity.

- Drawn lines: dispersion relation for gravity–capillary waves.

- Dashed lines: dispersion relation for deep-water gravity waves.

- Dash-dotted lines: dispersion relation valid for deep-water capillary waves.

In general, waves are also affected by gravity and are then called gravity–capillary waves. Their dispersion relation reads, for waves on the interface between two fluids of infinite depth:

$$\omega^2 = |k|\left(\frac{\rho - \rho'}{\rho + \rho'}g + \frac{\sigma}{\rho + \rho'}k^2\right),$$

Where g is the acceleration due to gravity, ρ and ρ' are the mass density of the two fluids ($\rho > \rho'$). The factor $(\rho - \rho')/(\rho + \rho')$ in the first term is the Atwood number.

Gravity Wave Regime

For large wavelengths (small $k = 2\pi/\lambda$), only the first term is relevant and one has gravity waves. In this limit, the waves have a group velocity half the phase velocity: following a single wave's crest in a group one can see the wave appearing at the back of the group, growing and finally disappearing at the front of the group.

Capillary Wave Regime

Shorter (large k) waves (e.g., 2 mm for the water–air interface), which are proper capillary waves, do the opposite: an individual wave appears at the front of the group, grows when moving towards the group center and finally disappears at the back of the group. Phase velocity is two thirds of group velocity in this limit.

Phase Velocity Minimum

Between these two limits is a point at which the dispersion caused by gravity cancels out the dispersion due to the capillary effect. At a certain wavelength, the group velocity equals the phase velocity, and there is no dispersion. At precisely this same wavelength, the phase velocity of gravity–capillary waves as a function of wavelength (or wave number) has a minimum. Waves with wavelengths much smaller than this critical wavelength λ_m are dominated by surface tension, and much above by gravity. The value of this wavelength and the associated minimum phase speed c_m are:

$$\lambda_m = 2\pi \sqrt{\frac{\sigma}{(\rho - \rho')g}} \quad \text{and} \quad c_m = \sqrt{\frac{2\sqrt{(\rho - \rho')g\sigma}}{\rho + \rho'}}.$$

For the air–water interface, λ_m is found to be 1.7 cm (0.67 in), and c_m is 0.23 m/s (0.75 ft/s).

If one drops a small stone or droplet into liquid, the waves then propagate outside an expanding circle of fluid at rest; this circle is a caustic which corresponds to the minimal group velocity.

Derivation

As Richard Feynman put it, "[water waves] that are easily seen by everyone and which are usually used as an example of waves in elementary courses are the worst possible example; they have all the complications that waves can have." The derivation of the general dispersion relation is therefore quite involved.

There are three contributions to the energy, due to gravity, to surface tension, and to hydrodynamics. The first two are potential energies, and responsible for the two terms inside the parenthesis, as is clear from the appearance of g and σ. For gravity, an assumption is made of the density of the fluids being constant (i.e., incompressibility), and likewise g (waves are not high enough for gravitation to change appreciably). For surface tension, the deviations from planarity (as measured by derivatives of the surface) are supposed to be small. For common waves both approximations are good enough.

The third contribution involves the kinetic energies of the fluids. It is the most complicated and calls for a hydrodynamic framework. Incompressibility is again involved (which is satisfied if the speed of the waves is much less than the speed of sound in the media), together with the flow being irrotational – the flow is then potential. These are typically also good approximations for common situations.

The resulting equation for the potential (which is Laplace equation) can be solved with the proper boundary conditions. On one hand, the velocity must vanish well below the surface (in the "deep water" case, which is the one we consider, otherwise a more involved result is obtained, see Ocean surface waves.) On the other, its vertical component must match the motion of the surface. This contribution ends up being responsible for the extra k outside the parenthesis, which causes all regimes to be dispersive, both at low values of k, and high ones (except around the one value at which the two dispersions cancel out).

Waves and Shallow Water

When waves travel into areas of shallow water, they begin to be affected by the ocean bottom. The free orbital motion of the water is disrupted, and water particles in orbital motion no longer return to their original position. As the water becomes shallower, the swell becomes higher and steeper, ultimately assuming the familiar sharp-crested wave shape. After the wave breaks, it becomes a wave of translation and erosion of the ocean bottom intensifies.

Shallow Water Equations

The shallow water equations are a set of hyperbolic partial differential equations (or parabolic if viscous shear is considered) that describe the flow below a pressure surface in a fluid (sometimes, but not necessarily, a free surface). The shallow water equations in unidirectional form are also called Saint-Venant equations.

The equations are derived from depth-integrating the Navier–Stokes equations, in the case where the horizontal length scale is much greater than the vertical length scale. Under this condition, conservation of mass implies that the vertical velocity scale of the fluid is small compared to the horizontal velocity scale. It can be shown from the momentum equation that vertical pressure gradients are nearly hydrostatic, and that horizontal pressure gradients are due to the displacement of the pressure surface, implying that the horizontal velocity field is constant throughout the depth of the fluid. Vertically integrating allows the vertical velocity to be removed from the equations. The shallow water equations are thus derived.

While a vertical velocity term is not present in the shallow water equations, note that this velocity is not necessarily zero. This is an important distinction because, for example, the vertical velocity cannot be zero when the floor changes depth, and thus if it were zero only flat floors would be usable with the shallow water equations. Once a solution (i.e. the horizontal velocities and free surface displacement) has been found, the vertical velocity can be recovered via the continuity equation.

Situations in fluid dynamics where the horizontal length scale is much greater than the vertical length scale are common, so the shallow water equations are widely applicable. They are used with Coriolis forces in atmospheric and oceanic modeling, as a simplification of the primitive equations of atmospheric flow.

Shallow water equation models have only one vertical level, so they cannot directly encompass any factor that varies with height. However, in cases where the mean state is sufficiently simple, the vertical variations can be separated from the horizontal and several sets of shallow water equations can describe the state.

Equations

Conservative Form

The shallow water equations are derived from equations of conservation of mass and conservation of linear momentum (the Navier–Stokes equations), which hold even when the assumptions of shallow water break down, such as across a hydraulic jump. In the case of a horizontal bed, no Coriolis forces, frictional or viscous forces, the shallow-water equations are:

$$\frac{\partial(\rho\eta)}{\partial t}+\frac{\partial(\rho\eta u)}{\partial x}+\frac{\partial(\rho\eta v)}{\partial y}=0,$$

$$\frac{\partial(\rho\eta u)}{\partial t}+\frac{\partial}{\partial x}\left(\rho\eta u^{2}+\frac{1}{2}\rho g\eta^{2}\right)+\frac{\partial(\rho\eta uv)}{\partial y}=0,$$

$$\frac{\partial(\rho\eta v)}{\partial t}+\frac{\partial(\rho\eta uv)}{\partial x}+\frac{\partial}{\partial y}\left(\rho\eta v^{2}+\frac{1}{2}\rho g\eta^{2}\right)=0.$$

Here η is the total fluid column height (instantaneous fluid depth as a function of x, y and t), and the 2D vector (u,v) is the fluid's horizontal flow velocity, averaged across the vertical column. Further g is acceleration due to gravity and ρ is the fluid density. The first equation is derived from mass conservation, the second two from momentum conservation.

Non-conservative Form

Expanding the derivatives in the above using the product rule, the non-conservative form of the shallow-water equations is obtained. Since velocities are not subject to a fundamental conservation equation, the non-conservative forms do not hold across a shock or hydraulic jump. Also included are the appropriate terms for Coriolis, frictional and viscous forces, to obtain (for constant fluid density):

$$\frac{\partial h}{\partial t}+\frac{\partial}{\partial x}\left((H+h)u\right)+\frac{\partial}{\partial y}\left((H+h)v\right)=0,$$

$$\frac{\partial u}{\partial t}+u\frac{\partial u}{\partial x}+v\frac{\partial u}{\partial y}-fv=-g\frac{\partial h}{\partial x}-bu+v\left(\frac{\partial^{2}u}{\partial x^{2}}+\frac{\partial^{2}u}{\partial y^{2}}\right),$$

$$\frac{\partial v}{\partial t}+u\frac{\partial v}{\partial x}+v\frac{\partial v}{\partial y}+fu=-g\frac{\partial h}{\partial y}-bv+v\left(\frac{\partial^{2}v}{\partial x^{2}}+\frac{\partial^{2}v}{\partial y^{2}}\right),$$

where,

- U is the velocity in the x direction, or zonal velocity.

- v is the velocity in the y direction, or meridional velocity.

- h is the height deviation of the horizontal pressure surface from its mean height H: $\eta = H + h$

- H is the mean height of the horizontal pressure surface.

- G is the acceleration due to gravity.

- F is the Coriolis coefficient associated with the Coriolis force. On Earth, f is equal to $2\Omega\sin(\varphi)$, where Ω is the angular rotation rate of the Earth ($\pi/12$ radians/hour), and φ is the latitude.

- B is the viscous drag coefficient.

- ν is the kinematic viscosity.

It is often the case that the terms quadratic in u and v, which represent the effect of bulk advection, are small compared to the other terms. This is called geostrophic balance, and is equivalent to saying that the Rossby number is small. Assuming also that the wave height is very small compared to the mean height ($h \ll H$), we have (without lateral viscous forces):

$$\frac{\partial h}{\partial t} + H\left(\frac{\partial u}{\partial x} + \frac{\partial v}{\partial y}\right) = 0,$$

$$\frac{\partial u}{\partial t} - fv = -g\frac{\partial h}{\partial x} - bu,$$

$$\frac{\partial v}{\partial t} + fu = -g\frac{\partial h}{\partial y} - bv.$$

One-dimensional Saint-venant Equations

The one-dimensional (1-D) Saint-Venant equations were derived by Adhémar Jean Claude Barré de Saint-Venant, and are commonly used to model transient open-channel flow and surface runoff. They can be viewed as a contraction of the two-dimensional (2-D) shallow water equations, which are also known as the two-dimensional Saint-Venant equations. The 2-D Saint-Venant equations contain to a certain extent the main characteristics of the channel cross-sectional shape.

The 1-D equations are used extensively in computer models such as Mascaret (EDF), SIC (Irstea), HEC-RAS, SWMM5, ISIS, InfoWorks, Flood Modeller, SOBEK 1DFlow, MIKE 11, and MIKE SHE because they are significantly easier to solve than the full shallow water equations. Common applications of the 1-D Saint-Venant equations include flood routing along rivers (including evaluation of measures to reduce the risks of flooding), dam break analysis, storm pulses in an open channel, as well as storm runoff in overland flow.

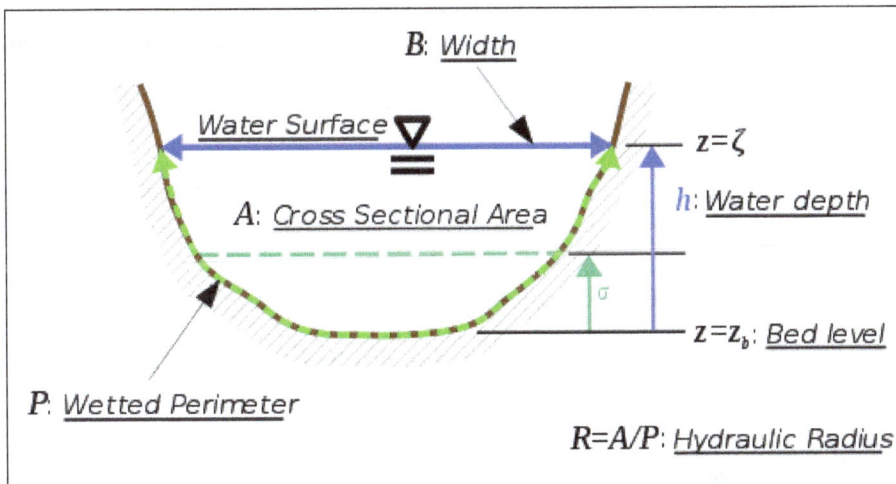

Cross section of the open channel.

The system of partial differential equations which describe the 1-D incompressible flow in an open channel of arbitrary cross section – as derived and posed by Saint-Venant:

$$\frac{\partial A}{\partial t} + \frac{\partial (Au)}{\partial x} = 0 \quad \text{and}$$

$$\frac{\partial u}{\partial t} + u\frac{\partial u}{\partial x} + g\frac{\partial \zeta}{\partial x} = -\frac{P}{A}\frac{\tau}{\rho},$$

where x is the space coordinate along the channel axis, t denotes time, $A(x, t)$ is the cross-sectional area of the flow at location x, $u(x, t)$ is the flow velocity, $\zeta(x, t)$ is the free surface elevation and $\tau(x, t)$ is the wall shear stress along the wetted perimeter $P(x, t)$ of the cross section at x. Further ρ is the (constant) fluid density and g is the gravitational acceleration.

Closure of the hyperbolic system of equations $\frac{\partial A}{\partial t} + \frac{\partial (Au)}{\partial x} = 0 \ -\frac{\partial u}{\partial t} + u\frac{\partial u}{\partial x} + g\frac{\partial \zeta}{\partial x} = -\frac{P}{A}\frac{\tau}{\rho}$, is obtained from the geometry of cross sections – by providing a functional relationship between the cross-sectional area A and the surface elevation ζ at each position x. For example, for a rectangular cross section, with constant channel width B and channel bed elevation z_b, the cross sectional area is: $A = B(\zeta - z_b) = B\,h$. The instantaneous water depth is $h(x, t) = \zeta(x, t) - z_b(x)$, with $z_b(x)$ the bed level (i.e. elevation of the lowest point in the bed above datum,). For non-moving channel walls the cross-sectional area A in equation $\frac{\partial A}{\partial t} + \frac{\partial (Au)}{\partial x} = 0$ can be written as:

$$A(x,t) = \int_0^{h(x,t)} b(x,h')\,\mathrm{d}h',$$

with $b(x, h)$ the effective width of the channel cross section at location x when the fluid depth is h – so $b(x,h) = B(x)$ for rectangular channels.

The wall shear stress τ is dependent on the flow velocity u, they can be related by using e.g. the Darcy–Weisbach equation, Manning formula or Chézy formula.

Further, equation $\frac{\partial A}{\partial t} + \frac{\partial (Au)}{\partial x} = 0$ is the continuity equation, expressing conservation of water volume for this incompressible homogeneous fluid. Equation

$$\frac{\partial u}{\partial t} + u\frac{\partial u}{\partial x} + g\frac{\partial \zeta}{\partial x} = -\frac{P}{A}\frac{\tau}{\rho},$$

is the momentum equation, giving the balance between forces and momentum change rates.

The bed slope $S(x)$, friction slope $S_f(x, t)$ and hydraulic radius $R(x, t)$ are defined as:

$$S = -\frac{\mathrm{d}z_b}{\mathrm{d}x}, \quad S_f = \frac{\tau}{\rho g R} \quad \text{and} \quad R = \frac{A}{P}.$$

Consequently, the momentum equation:

$$\frac{\partial u}{\partial t}+u\frac{\partial u}{\partial x}+g\frac{\partial \zeta}{\partial x}=-\frac{P}{A}\frac{\tau}{\rho},$$

can be written as:

$$\frac{\partial u}{\partial t}+u\frac{\partial u}{\partial x}+g\frac{\partial h}{\partial x}+g\left(S_f-S\right)=0.$$

Conservation of Momentum

The momentum equation:

$$\frac{\partial u}{\partial t}+u\frac{\partial u}{\partial x}+g\frac{\partial h}{\partial x}+g\left(S_f-S\right)=0.$$

can also be cast in the so-called conservation form, through some algebraic manipulations on the Saint-Venant equations, $\frac{\partial A}{\partial t}+\frac{\partial (Au)}{\partial x}=0$ and $\frac{\partial u}{\partial t}+u\frac{\partial u}{\partial x}+g\frac{\partial h}{\partial x}+g\left(S_f-S\right)=0.$ In terms of the discharge $Q = Au$:

$$\frac{\partial Q}{\partial t}+\frac{\partial}{\partial x}\left(\frac{Q^2}{A}+g\,I_1\right)+g\,A\left(S_f-S\right)-g\,I_2=0,$$

where A, I_1 and I_2 are functions of the channel geometry, described in the terms of the channel width $B(\sigma,x)$. Here σ is the height above the lowest point in the cross section at location x, So σ is the height above the bed level $z_b(x)$ (of the lowest point in the cross section):

$$A(\sigma,x)=\int_0^\sigma B(\sigma',x)\,d\sigma',$$

$$I_1(\sigma,x)=\int_0^\sigma(\sigma-\sigma')B(\sigma',x)\,d\sigma'$$

$$I_2(\sigma,x)=\int_0^\sigma(\sigma-\sigma')\frac{\partial B(\sigma',x)}{\partial x}\,d\sigma'.$$

Above in the momentum equation:

$$\frac{\partial Q}{\partial t}+\frac{\partial}{\partial x}\left(\frac{Q^2}{A}+g\,I_1\right)+g\,A\left(S_f-S\right)-g\,I_2=0,$$

in conservation form – A, I_1 and I_2 are evaluated at $\sigma = h(x, t)$. The term $g\,I_1$ describes the hydrostatic force in a certain cross section. And, for a non-prismatic channel, $g\,I_2$ gives the effects of geometry variations along the channel axis x.

In applications, depending on the problem at hand, there often is a preference for using either the momentum equation in non-conservation form,

$$\frac{\partial u}{\partial t} + u\frac{\partial u}{\partial x} + g\frac{\partial \zeta}{\partial x} = -\frac{P}{A}\frac{\tau}{\rho},$$

or

$$\frac{\partial u}{\partial t} + u\frac{\partial u}{\partial x} + g\frac{\partial h}{\partial x} + g\left(S_{\mathrm{f}} - S\right) = 0.$$

The conservation form:

$$\frac{\partial Q}{\partial t} + \frac{\partial}{\partial x}\left(\frac{Q^2}{A} + g\,I_1\right) + g\,A\left(S_f - S\right) - g\,I_2 = 0,$$

For instance in case of the description of hydraulic jumps, the conservation form is preferred since the momentum flux is continuous across the jump.

Characteristics

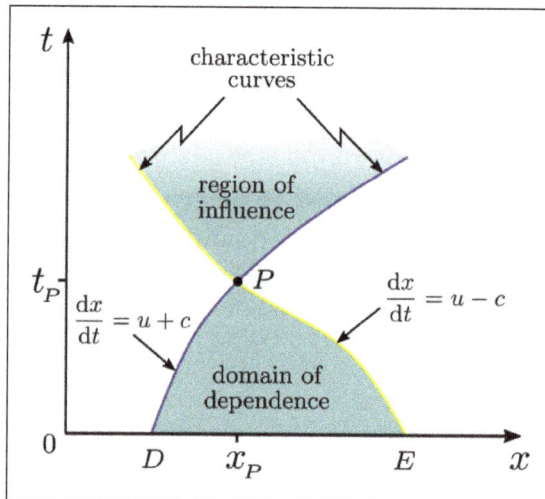

Characteristics, domain of dependence and region of influence, associated with location $P = (x_p, t_p)$ in space x and time t.

The Saint-Venant equations:

$$\frac{\partial A}{\partial t} + \frac{\partial\left(Au\right)}{\partial x} = 0 \quad - \quad \frac{\partial u}{\partial t} + u\frac{\partial u}{\partial x} + g\frac{\partial \zeta}{\partial x} = -\frac{P}{A}\frac{\tau}{\rho},$$

can be analysed using the method of characteristics. The two celerities dx/dt on the characteristic curves are:

$$\frac{dx}{dt} = u \pm c, \quad \text{with} \quad c = \sqrt{\frac{gA}{B}}.$$

The Froude number $F = |u|/c$ determines whether the flow is subcritical ($F < 1$) or supercritical ($F > 1$). For a rectangular and prismatic channel of constant width B, i.e. with $A = B\,h$ and $c = \sqrt{gh}$, the Riemann invariants are:

$$r_+ = u + 2\sqrt{gh} \quad and \quad r_- = u - 2\sqrt{gh},$$

so the equations in characteristic form are:

$$\frac{d}{dt}\left(u + 2\sqrt{gh}\right) = g\left(S - S_f\right) \text{along} \quad \frac{dx}{dt} = u + \sqrt{gh} \quad and$$

$$\frac{d}{dt}\left(u - 2\sqrt{gh}\right) = g\left(S - S_f\right) \text{along} \quad \frac{dx}{dt} = u - \sqrt{gh}.$$

The Riemann invariants and method of characteristics for a prismatic channel of arbitrary cross-section are described by Didenkulova & Pelinovsky.

The characteristics and Riemann invariants provide important information on the behavior of the flow, as well as that they may be used in the process of obtaining (analytical or numerical) solutions.

Derived Modelling

Dynamic Wave

The dynamic wave is the full one-dimensional Saint-Venant equation. It is numerically challenging to solve, but is valid for all channel flow scenarios. The dynamic wave is used for modeling transient storms in modeling programs including Mascaret (EDF), SIC (Irstea), HEC-RAS, InfoWorks_ICM, MIKE 11, Wash 123d and SWMM5.

In the order of increasing simplifications, by removing some terms of the full 1D Saint-Venant equations (aka Dynamic wave equation), we get the also classical Diffusive wave equation and Kinematic wave equation.

Diffusive Wave

For the diffusive wave it is assumed that the inertial terms are less than the gravity, friction, and pressure terms. The diffusive wave can therefore be more accurately described as a non-inertia wave, and is written as:

$$g\frac{\partial h}{\partial x} + g(S_f - S) = 0.$$

The diffusive wave is valid when the inertial acceleration is much smaller than all other forms of acceleration, or in other words when there is primarily subcritical flow, with low Froude values.

Models that use the diffusive wave assumption include MIKE SHE and LISFLOOD-FP. In the SIC (Irstea) software this options is also available, since the 2 inertia terms (or any of them) can be removed in option from the interface.

Kinematic Wave

For the kinematic wave it is assumed that the flow is uniform, and that the friction slope is approximately equal to the slope of the channel. This simplifies the full Saint-Venant equation to the kinematic wave:

$$S_e - S = 0.$$

The kinematic wave is valid when the change in wave height over distance and velocity over distance and time is negligible relative to the bed slope, e.g. for shallow flows over steep slopes. The kinematic wave is used in HEC-HMS.

Derivation From Navier–stokes Equations

The 1-D Saint-Venant momentum equation can be derived from the Navier–Stokes equations that describe fluid motion. The x-component of the Navier–Stokes equations – when expressed in Cartesian coordinates in the x-direction – can be written as:

$$\frac{\partial u}{\partial t} + u\frac{\partial u}{\partial x} + v\frac{\partial u}{\partial y} + w\frac{\partial u}{\partial z} = -\frac{\partial p}{\partial x}\frac{1}{\rho} + v\left(\frac{\partial^2 u}{\partial x^2} + \frac{\partial^2 u}{\partial y^2} + \frac{\partial^2 u}{\partial z^2}\right) + f_x,$$

Where u is the velocity in the x-direction, v is the velocity in the y-direction, w is the velocity in the z-direction, t is time, p is the pressure, ρ is the density of water, v is the kinematic viscosity, and f_x is the body force in the x-direction.

1. If it is assumed that friction is taken into account as a body force, then v can be assumed as zero so:

$$v\left(\frac{\partial^2 u}{\partial x^2} + \frac{\partial^2 u}{\partial y^2} + \frac{\partial^2 u}{\partial z^2}\right) = 0.$$

2. Assuming one-dimensional flow in the x-direction it follows that:

$$v\frac{\partial u}{\partial y} + w\frac{\partial u}{\partial z} = 0$$

3. Assuming also that the pressure distribution is approximately hydrostatic it follows that:

$$p = \rho g h$$

or in differential form:

$$\partial p = \rho g(\partial h)$$

And when these assumptions are applied to the x-component of the Navier–Stokes equations:

$$-\frac{\partial p}{\partial x}\frac{1}{\rho} = -\frac{1}{\rho}\frac{\rho g(\partial h)}{\partial x} = -g\frac{\partial h}{\partial x}.$$

4. There are 2 body forces acting on the channel fluid, namely, gravity and friction:

$$f_x = f_{x,g} + f_{x,f}$$

where $f_{x,g}$ is the body force due to gravity and $f_{x,f}$ is the body force due to friction.

5. $f_{x,g}$ can be calculated using basic physics and trigonometry:

$$F_g = (\sin\theta)gM$$

Where F_g is the force of gravity in the x-direction, θ is the angle, and M is the mass.

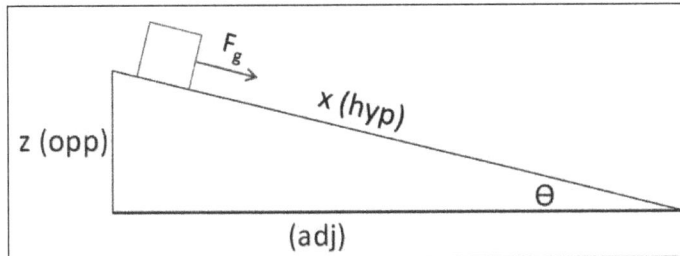

Diagram of block moving down an inclined plane.

The expression for sin θ can be simplified using trigonometry as:

$$\sin\theta = \frac{\text{opp}}{\text{hyp}}.$$

For small θ (reasonable for almost all streams) it can be assumed that:

$$\sin\theta = \tan\theta = \frac{\text{opp}}{\text{adj}} = S$$

And given that f_x represents a force per unit mass, the expression becomes:

$$f_{x,g} = gS.$$

6. Assuming the energy grade line is not the same as the channel slope, and for a reach of consistent slope there is a consistent friction loss, it follows that:

$$f_{x,f} = S_f g.$$

7. All of these assumptions combined arrives at the 1-dimensional Saint-Venant equation in the x-direction:

$$\frac{\partial u}{\partial t} + u\frac{\partial u}{\partial x} + g\frac{\partial h}{\partial x} + g(S_f - S) = 0,$$
$$(a) \quad (b) \quad (c) \quad\quad (d) \quad (e)$$

where (a) is the local acceleration term, (b) is the convective acceleration term, (c) is the pressure gradient term, (d) is the friction term, and (e) is the gravity term.

Terms

The local acceleration (a) can also be thought of as the "unsteady term" as this describes some change in velocity over time. The convective acceleration (b) is an acceleration caused by some change in velocity over position, for example the speeding up or slowing down of a fluid entering a constriction or an opening, respectively. Both these terms make up the inertia terms of the 1-dimensional Saint-Venant equation.

The pressure gradient term (c) describes how pressure changes with position, and since the pressure is assumed hydrostatic, this is the change in head over position. The friction term (d) accounts for losses in energy due to friction, while the gravity term (e) is the acceleration due to bed slope.

Wave Modelling by Shallow Water Equations

Shallow water equations can be used to model Rossby and Kelvin waves in the atmosphere, rivers, lakes and oceans as well as gravity waves in a smaller domain (e.g. surface waves in a bath). In order for shallow water equations to be valid, the wavelength of the phenomenon they are supposed to model has to be much larger than the depth of the basin where the phenomenon takes place. Somewhat smaller wavelengths can be handled by extending the shallow water equations using the Boussinesq approximation to incorporate dispersion effects. Shallow water equations are especially suitable to model tides which have very large length scales (over hundred of kilometers). For tidal motion, even a very deep ocean may be considered as shallow as its depth will always be much smaller than the tidal wavelength.

Turbulence Modelling using Non-linear Shallow Water Equations

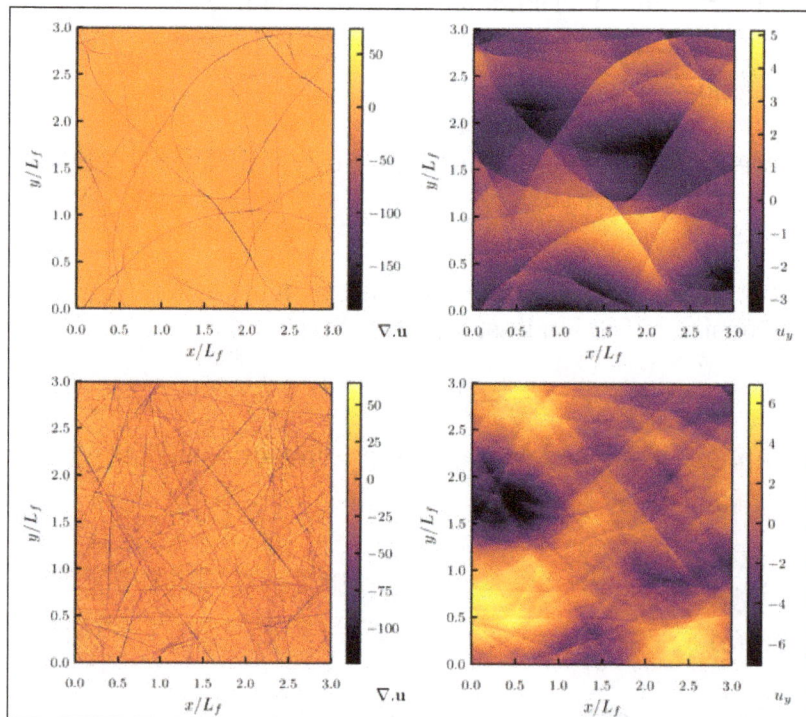

A snapshot from simulation of shallow water
equations in which shock waves are present.

Shallow water equations, in its non-linear form, is an obvious candidate for modelling turbulence in the atmosphere and oceans, i.e. geophysical turbulence. An advantage of this, over Quasi-geostrophic equations, is that it allows solutions like gravity waves, while also conserving energy and potential vorticity. However there are also some disadvantages as far as geophysical applications are concerned - it has a non-quadratic expression for total energy and a tendency for waves to become shock waves. Some alternate models have been proposed which prevent shock formation. One alternative is to modify the "pressure term" in the momentum equation, but it results in a complicated expression for kinetic energy. Another option is to modify the non-linear terms in all equations, which gives a quadratic expression for kinetic energy, avoids shock formation, but conserves only linearized potential vorticity.

Ocean Sediments

Sediment on the seafloor originates from a variety of sources, including biota from the overlying ocean water, eroded material from land transported to the ocean by rivers or wind, ash from volcanoes, and chemical precipitates derived directly from sea water. A very small amount of it even originates as interstellar dust. In short, the particles found in sediment on the seafloor vary considerably in composition and record a complex interplay of processes that have acted to form, transport, and preserve them.

Geological oceanographers have coined the terms "terrigenous" to describe those sediments derived from eroded material on land, "biogenic" for those derived from biological matter, "volcanogenic" for those that include significant amounts of ash, "hydrogenous" for those that precipitate directly from sea water, and "cosmogenic" for those that come from interstellar space.

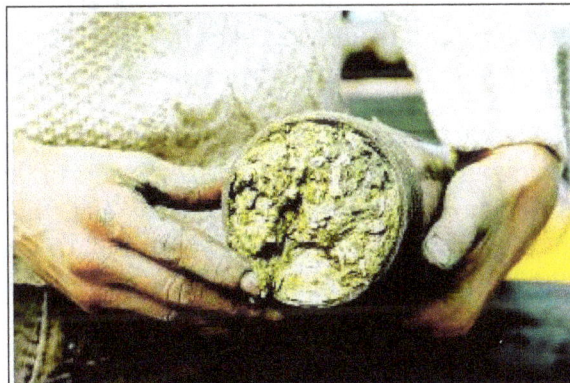

A core sample of sediment from Chesapeake Bay can tell scientists about
the oceanographic history of that particular location, including climate
change, pollution, and past changes in erosion.

The seafloor, however, is not a random arrangement of these different sediment types. Oceanographers have painstakingly mapped the distribution of sediment around the globe and have learned that at any given location the sediments provide important information regarding the history of the ocean as well as the overall state of climate on the Earth's surface. By studying how the heterogeneous composition of sediment varies as a function of geographic location and age, oceanographers are able to document the geologic and climatic conditions that are responsible for that sediment.

Oceanographers study sediment by taking long cylindrical cores, which individually can be as long as 18 to 30 meters (60 to 98 feet). Because the bottom of the ocean is extremely cold (only 1 to 3 degrees above freezing), the cores are stored in refrigerators onboard the research ship prior to being stored in large refrigerated repositories at shore-based laboratories. In their laboratories, scientists study the physical, chemical, and biological makeup of the sediment.

Regardless of which type of sediment, there are three processes that are responsible for its final composition: namely, the production of the sediment; its transport; and its preservation. It is important to differentiate between these three processes. For example, if a sedimentary particle is produced, but not preserved, there will be no resulting sedimentary record. Thus, only if material is produced and transported and preserved will marine sediment result.

The different combinations of each process' effectiveness result in a commensurate variety of sedimentation rates. Sediment can accumulate as slowly as 0.1 millimeter (0.04 inch) per 1,000 years (in the middle of the ocean where only wind-blown material is deposited) to as fast as 1 meter (3.25 feet) per year along continental margins. More typical deep-sea rates are on the order of several centimeters per 1,000 years.

Production of Sediment

The production of marine sediment is more complex than it may seem. Terrigenous sediment is produced by an interplay of chemical and physical weathering processes, which collectively serve to create small grains of material ranging in size from thousandths of millimeters to 1 or 2 millimeters (0.04 or 0.08 inch).

Physical weathering is caused by mechanical fracturing of rocks, such as that due to the freezing of water in cracks, and results in finer grained, compositionally similar examples of the original rock. On the other hand, chemical weathering, caused by the weak acid produced by the interaction of rainwater and atmospheric carbon dioxide, degrades the rock slowly and often produces fine-grained minerals that are compositionally distinct from the original rock.

Biogenic (biologically derived) sediment is produced by marine plankton, which are small, often microscopic, unicellular plants and animals that float in the surface waters of the ocean. The shells of these organisms are made of either calcium carbonate ($CaCO_3$) or silica (SiO_2). Although ubiquitous, particularly elevated concentrations of such organisms are most commonly found in biologically productive waters such as the Equatorial Pacific, or the Southern Ocean ringing the continent of Antarctica.

Volcanic ash is produced during volcanic eruptions, as can be seen in the billowing ejected material from many volcanos. Cosmogenic material is the remains of primordial material left over from the creation of the solar system (and perhaps from beyond) and, although very low in abundance, is ubiquitously distributed.

The production of hydrogenous sediment is most difficult to visualize, but involves either the slow precipitation of dissolved chemicals from sea water or the leaching of chemical elements from rocks that have extremely hot sea water (greater than 300 °C [572 °F]) circulating through them along mid-ocean ridges. When these hot solutions are injected into the cold sea water the leached chemical elements precipitate from the cooling water, leading to hydrothermal sediments near the mid-ocean ridge that are enriched in iron, manganese, copper, zinc, and other metals.

Transport of Sediment

The transport of sediment depends on its grain size and the original location where it was produced. Terrigenous sediment can be transported to the deep sea via rivers or by wind. Material transported by rivers most commonly ends up deposited on the continental margin, the shallow portions of the ocean that are within several hundred kilometers of land. When continental margin deposits accumulate fast and get overly steep, or when an earthquake or storm causes the sediment to be resuspended, turbidity currents provide additional transport out to the deep sea. The resuspension of the sediment into the bottom water causes it to be more dense than the overlying water, and thus these turbidity currents flow downslope to the more distant ocean basin.

The transport of sediment by wind is also extremely significant, and is particularly relevant to studies of Earth's climate in the past. When the Earth's climate is relatively dry (arid), such as during glacial periods, the land surface tends to be more dusty, and thus during such periods there will be more windblown terrigenous material delivered to the deep ocean. Also, during such time periods the wind speed tends to be higher, and thus terrigenous grains that are slightly larger than usual are preferentially transported. Thus, by examining the amount of dust, as well as its grain size, in the different layers of a sediment core, oceanographers learn how arid the land surface was at a given time, as well as how fast the average wind speeds were.

Although such dust is essentially invisible to the human eye, its transport is still an important and long-ranging process. For example, dust derived from the Sahara in North Africa is easily observed in Miami, Florida and even in the eastern Pacific Ocean. Moreover, volcanic ash ejected tens of kilometers into the atmosphere during the largest eruptions can be transported by winds all around the globe.

Wind-swept desert sands not only produce a cooling effect due to deflection of incoming solar radiation, but they also deposit sand, silt and dust on the ocean surface and ultimately on the ocean floor. Through mineralogical and chemical analysis, scientists can recreate historical patterns in climate and geological development.

The microscopic shells of the plankton do not just simply fall to the seafloor. In fact, because they are so small, the plankton may not be able to fall individually. Oceanographers learn how such sediment is delivered to the seafloor by suspending sediment traps in the ocean. These traps are essentially large funnels, up to 1 or 2 meters (3.3 to 6.6 feet) in diameter, that collect the material as it falls through sea water.

By examining the material trapped by these instruments, it was discovered that plankton shells are delivered to seafloor by "biopackaging" via fecal pellets. In other words, various microorganisms that eat other plankton excrete their shells in fecal pellets. These "biopackaged" fecal pellets are large enough (0.2-1.5 millimeters, or 0.008-0.059 inch) and dense enough to sink to the seafloor, where they become part of the sediment.

Hydrothermal sediment is largely localized to within less than 10 kilometers (6.2 miles) of the mid-ocean ridge. The concentration of metals in these sediments decreases with distance from a ridge, yet small amounts can be found up to 500 to 1000 kilometers (300-600 miles) away.

Preservation of Sediment

Terrigenous sediment, whether it be delivered by rivers or wind, is not altered significantly on the seafloor and thus is well-preserved. During very deep burial (e.g., 5 kilometers, or 3 miles, below the seafloor), the terrigenous grains can be altered into different minerals, but this does not occur while the grains are lying on the seafloor and is generally a more important process for geologists rather than oceanographers.

Biogenic sediment, on the other hand, is very poorly preserved on the seafloor. The degradation of biogenic sediments is a complex, largely chemical suite of processes. The preservation of these sediments is a field of study that has captivated oceanographers for over 100 years, dating from when they were discovered in the mid-1800s during the first oceanographic research cruise by the ship HMS Challenger.

For example, significantly less than 1 percent of the siliceous plankton that are biopackaged to the seafloor are preserved. This is because sea water is undersaturated with respect to silica. Therefore, the siliceous plankton are living in an environment that is corrosive to their shells.

While the plankon is alive, the shell is surrounded by organic protoplasm that protects it from the corrosive sea water. After death, however, even if biopackaged, this organic coating will degrade, exposing the shell of the siliceous plankton. When exposed to sea water, the shell will dissolve.

This process occurs over all depth and temperature ranges throughout the global ocean. Thus, the only regions of the seafloor where biogenic silica appreciably accumulates is where the production of biogenic silica is so enormous that it overwhelms the amount that is dissolved. In the modern oceans, this occurs at high latitudes in the North Pacific and Southern Ocean and the Equatorial Pacific Ocean.

Plankton with shells made of calcium carbonate also commonly dissolve, but not as commonly as siliceous plankton. The dissolution of carbonate plankton is controlled by water depth and water temperature. Water depth and hydrostatic pressure correlate with each other—at greater depths there is greater pressure. At greater pressures, the solubility of carbon dioxide gas increases. An excellent analogy of this process is observed in a bottled carbonated beverage that is under pressure until opened—when the pressure is released the carbon dioxide comes out of solution and bubbles form. Similarly, at the great depths of the deepest seafloor, the solubility of carbon dioxide increases so much that calcium carbonate sediment may dissolve. This dissolution is also facilitated at the lower temperatures of the deep sea.

The converse is also true. At shallow water depths (that is, lower pressure) the carbonate does not dissolve and the warmer water temperatures (along with the increased light for photosynthesis) each serve to enhance the construction and preservation of coral reefs and other carbonate-producing biota. Thus, there is both a depth and latitudinal effect on the distribution of carbonate sediments due to their influence on temperature and pressure.

Sediments and Biogeochemical Cycles

Sediments cover most of the seafloor and are important as sinks and sources in the biogeochemical cycles of elements in seawater.

Classification of Sediments

All sediments are mixtures of particles from different sources and of different composition. However, they may be classified by their general characteristics.

Classification by Grain Size

Sediments may be classified by grain size, Sediment particles range from boulders to grains so small that they cannot be distinguished except under the most powerful microscopes. However, most sediments deposited at a specific location and time consist primarily of grains that are within a narrow range of particle sizes and are said to be well sorted.

Sediments are generally classified as gravel, sand, or mud (which can be subdivided into silt and clay) in descending order of their predominant grain size.

Classification by Origin

Sediments may be Classified by the Origin of the Majority of their Particles

Sediment particles may be derived from land (lithogenous), biological processes (biogenous), chemical precipitation from sweater (hydrogenous), or derived from meteorites (cosmogenous).

The distribution of sediment particles from these different origins within ocean sediments at different locations is determined by many factors including grain size, location of origin, susceptibility to decomposition or dissolution in seawater, and the mechanisms of particle transport.

Lithogenous Sediments

Lithogenous sediment particles are produced by wind and water erosion and weathering of terrestrial rocks. During weathering, easily dissolved minerals are removed, leaving mostly siliceous minerals including quartz, feldspars, and clay minerals. Clay minerals are layered structures of silicon, aluminum, and oxygen atoms, some containing iron and other elements. They are carried to the oceans by rivers, glaciers, and winds, or eroded from coastlines by waves.

Transport by Rivers

Most rivers slow down as they near the ocean, and many flow slowly over flat coastal plains so that larger particles are generally deposited in river valleys and only small particles are carried out to the oceans. However, during storm runoff events large amounts of larger particles can be resuspended from the river bed and carried to the ocean, where they accumulate in sediments near the river mouth.

Rivers that flow across active subduction zone margins are generally short due to the coastal mountain ranges and carry relatively small quantities of suspended sediment because they drain only small area of land.

Approximately 90% of all lithogenous sediments reach the oceans through rivers and 80% of this input is derived from Asia. The largest amounts are carried by four rivers: the Ganges, Brahmaputra, and the Irrawaddy empty into the Bay of Bengal, and the Indus discharges into the Arabian Gulf. In the area where these rivers discharge to the oceans extensive deep sediment deposits extend to the deep sea floor in the form of abyssal fans.

Most of the other rivers that transport large quantities of suspended sediments to the oceans empty into marginal seas.

Erosion by Glaciers

Glaciers erode very large amount of rock from their valleys, producing particles ranging from large boulders to extremely fine particles. The eroded rock is bulldozed, dragged, and carried down the glacier's valley and deposited at its lower end.

Many high-latitude glaciers empty directly into the oceans or end near where the glacial valley meets the sea. Icebergs that break off glaciers can carry their eroded rock long distances into the oceans to be released as the iceberg melts

Glaciers release their eroded rock at their lower ends. Much of this material then washes out to the ocean, especially the very fine-grained material called glacial flour (because it remains suspended for long periods and can give the water a milky appearance in lakes or fjords where the glacier empties).

Erosion by Waves

Waves continuously erode coastlines, Sediment particles released to the oceans by coastal wave erosion are similar to riverborne suspended sediments but often have a larger proportion of un-weathered mineral grains. Wave erosion creates particles of all sizes. Wave action then sorts the particles, transporting small ones offshore and leaving larger ones on or close to the shore.

Transport by Winds

Dust particles can be carried very long distances through the atmosphere. For example, Sahara Desert sand grain dust is transported across the Atlantic Ocean, where it can be easily identified on air filters placed at the coastline in Florida.

Dust particles in the atmosphere are eventually deposited on the ocean surface and sink slowly to the seafloor. Fine wind-blown dust particles fall over all parts of the oceans at a relatively uniform rate. Although this rate is slow, atmospheric dust accumulate continuously and is a major component of sediments that are remote from land and have very slow accumulation rates of other types of particles.

Explosive volcanic eruptions of subduction zone volcanoes can inject very large quantities of fragmented rock into the atmosphere. Historical records suggest that large eruptions such as the one that created Long Valley caldera in California can inject hundreds of cubic kilometers of pulverized rock into the atmosphere.

The largest particles injected into the atmosphere by volcanic eruptions rain out near the eruption site, but smaller particles can stay in the atmosphere for years and are distributed around the world before they eventually rain out onto the land or ocean.

Transport by Landslides

Landslides on coastlines can carry lithogenous materials of a wide range of particle sizes into the oceans, where they are incorporated in the sediments.

Slumps and turbidity currents on the continental slopes can carry lithogenous sediments from the continental shelf into trenches or out over the abyssal plain where there is no trench.

Biogenous Sediments

Almost all ocean life depends on photosynthesis performed by microscopic organism called phytoplankton. Most phytoplankton are consumed by larger organisms called zooplankton that excrete organic-rich fecal material. This material is often in the form of fecal pellets, which are larger than the individual phytoplankton that compose them and sink much faster. Marine species larger than zooplankton also produce fecal pellets.

Much of the organic matter in fecal pellets is decomposed by bacteria and other organisms or consumed by detritus feeders as it sinks through the water column or after it has been deposited on the seafloor. As a result, most ocean sediments contain little organic matter. Biogenous particles are predominantly the solution-resistant silica or calcium carbonate hard parts of microscopic phytoplankton and zooplankton.

The two major factors affecting the accumulation rate of biogenous particles in sediments are the rate of production of the particles in the overlying water column and the rate of decomposition or dissolution of the particles.

Biogenous particles can dominate sediments in areas where the productivity is high in the overlying water, but not where the hard parts are dissolved before they can accumulate in the sediments.

Regional Variations of Biogenous Particle Production

In high latitudes diatoms are the dominant photosynthetic organisms. They have siliceous hard parts and they dominate the inputs of biogenous material to the sediments.

At lower latitudes, many of the dominant photosynthetic organisms have no hard parts so inputs of biogenous material to sediments are limited except in certain regions where coccolithophores grow in abundance. Coccolithophores have calcium carbonate hard parts that can become the dominant particle in sediments beneath areas of high coccolithophore productivity if the water depth is not great enough so that the calcium carbonate is dissolved.

Some zooplankton, small free-floating animals that eat phytoplankton, also have hard parts that can contribute to, or dominate, sediments in some areas. Foraminifera and pteropods have calcareous hard parts, whereas radiolaria have silica hard parts.

Radiolaria are abundant in those tropical waters that have high productivity and can dominate sediments in these areas.

Dissolution of Biogenous Particles

Calcium carbonate particles dissolve more easily at higher pressures (depths), whereas silica particles dissolve very slowly at all depths and their dissolution rate actually decreases with increasing depth.

Thin siliceous hard parts may be dissolved before they reach the seafloor but not thicker hard parts. For example, hard parts from diatoms will accumulate in the sediments whatever the depth if they are abundant in the overlying water.

The upper layers of seawater are generally saturated or supersaturated with calcium carbonate, so calcareous material does not dissolve. However, calcium carbonate solubility increases with increasing pressure (depth) and decreasing temperature. Because deep water in the oceans is cold, the calcium carbonate dissolution rate increases substantially with increasing depth.

There are two types of calcium carbonate hard parts: calcite and aragonite. Some types of animal have calcite hard parts (e.g., pteropods) and others aragonite (e.g., foraminifera). Aragonite dissolves more easily than calcite so pteropod hard parts are totally dissolved at shallower depths than foraminifera hard parts. Where pteropods are more abundant than foraminifera, sediments at shallow depths may consist primarily of pteropod remains, in deeper water the pteropods are dissolved and sediments may be dominated by foraminifera remains, and at even deeper depths both forms of calcium carbonate are totally dissolved and the sediments contain no calcium carbonate.

Carbonate Compensation Depth

The depth below which all calcium hard parts are dissolved before they can accumulate in sediments is called the carbonate compensation depth, or CCD. No calcium carbonate–containing particles survive to be accumulated in surface sediments below the CCD.

Deep waters of the oceans are formed by sinking of cold surface water in certain regions near the poles. The cold water is saturated with carbon dioxide but as pressure increases the saturation solubility increases and deep water dissolves additional carbon dioxide released by respiration and decomposition of organic matter as the water flows through the ocean basins.

The CCD is affected not only by changes in temperature and pressure but also by changes in dissolved carbon dioxide concentration. Adding dissolved carbon dioxide to water lowers the pH and increases the solubility of calcium carbonate. As a result, the CCD is shallower in those area of the oceans where the deep water is older (a longer period of time since it left the surface). The CCD is shallower in the Pacific Ocean than in the Atlantic Ocean and is shallower in the South Atlantic Ocean than in the North Atlantic Ocean, reflecting the formation of deep water in an area near Greenland and the flow of this water southward to the south polar region and then around Antarctica and north into the Pacific Ocean. No deep water is formed in the North Pacific Ocean.

Carbonate Compensation Depth and the Greenhouse Effect

Changes in the mean CCD over time can have a major effect on the distribution of carbon dioxide between the atmosphere, oceans, and ocean sediments and so are important for greenhouse effect studies.

The higher concentration of carbon dioxide in surface ocean waters that has resulted from human use of fossil fuels will eventually reach the deep waters and may increase pH sufficiently to significantly reduce the CCD. This would cause more calcium carbonate to dissolve and the excess carbon dioxide could be released to the atmosphere when the water returns to the surface, providing a positive feedback to the enhanced greenhouse effect.

Hydrogenous Sediments

Hydrogenous sediments are precipitated from seawater predominantly as manganese and phosphorite nodules in certain areas near hydrothermal vents and in certain shallow tropical areas where conditions permit calcium carbonate to precipitate.

Hydrothermal Minerals

Heat flowing up through thin mantle, especially at the oceanic ridges, drives hydrothermal vents that discharge hot water. The mechanism involved is not known but is thought to be convection of water heated within the rocks and sediments and replacement of this water by percolation of seawater into the rocks and sediments from areas surrounding the vents.

Hydrothermal vents have been found on the oceanic ridges in many locations throughout the oceans and are probably quite common. The vents support communities of unique organisms that depend on chemosynthetic primary production by microbial organisms in the vent water as the ultimate source of their food.

The temperature and composition of the water discharged by hydrothermal vents varies. However, most vent plumes have no oxygen, substantial sulfide concentrations, and high concentrations of iron, manganese, and other metals (including copper, cobalt, lead, nickel, silver, zinc) that have soluble sulfides.

Once discharged into the surrounding ocean water, metal sulfides are oxidized and precipitate as a rain of fine particles of their hydrous oxides. Some particles sink to the seafloor to form metal-rich sediments in areas surrounding the vents and others are transported and deposited far from the vent to contribute to sediments elsewhere. Test-mining of metal-rich hydrothermal sediments has occurred in the Red Sea, where restricted circulation has allowed large concentrations of such sediments to accumulate.

Undersea Volcano Emissions

Hydrothermal vents have recently been found to exist on the flanks of hot-spot and magmatic-arc volcanoes where, in some instances, they discharge their fluids at much shallower depths (even sometimes within the photic zone) than at the oceanic ridges.

Manganese Nodules

Manganese nodules are dark brown, rounded lumps of rock, often larger than a potato, that are found lying on the abyssal ocean floor in high abundance in some locations.

Manganese nodules form by precipitation of minerals from seawater and are usually formed initially around a large sediment particle such as a shark tooth. The minerals build in concentric layers around the nodule in a manner similar to tree rings, and occasional disturbance by marine organisms is thought to be necessary for the nodules to grow on all sides and in order that they not be buried by new sediments.

Manganese nodules consist mostly of manganese oxide and iron oxide but also have high concentrations of other metals, including copper, nickel and zinc. The source of these minerals is not known but may be particles transported from hydrothermal vents.

Manganese nodules are potentially commercially valuable, especially in the central Pacific Ocean where they are most abundant.

Phosphorite Nodules and Crusts

Phosphorite nodules containing up to 30% phosphorus form in limited areas of the continental slope and some seamounts. They grow very slowly (1–10mm per 1000 yr) and their formation apparently requires low dissolved oxygen concentrations in the overlying bottom water and a large supply of phosphorus carried to the sediments by sinking detritus as a result of high productivity in the overlying surface waters.

Phosphorite nodules grow only by accumulation on their underside, where phosphorus is released by decomposition of detritus in the sediment.

Carbonates

Many limestone rocks lack fossils. In some limestone rocks, the fossils have been decomposed, but some other limestone rocks consist of calcium carbonate precipitated directly from seawater. Conditions that permit calcium carbonate precipitation must have been widespread in the past but are now found in very limited regions, such as the Bahamas.

Both high temperature and high productivity appear to be necessary for calcium carbonate precipitation to occur, as these conditions cause the pH to rise. Calcium carbonate is precipitated around suspended sediment particles to form rounded grains called ooliths.

Evaporites

In marginal seas with arid climates, evaporation may increase salinity so high that salts precipitate progressively from the seawater—first calcium and magnesium carbonate, then calcium sulfate, and finally sodium chloride.

Evaporites form in very few areas today but evaporite formation must have been more common at times in the past. For example, the Mediterranean Sea has several layers of evaporite sediments, some more than 100 m thick, indicating that the Mediterranean may have evaporated almost to dryness several times when sealevel fell and the connection with the Atlantic Ocean was broken.

Cosmogenous Sediments

Cosmogenous sediment particles are derived from meteors and meteorites and are relatively rare, although they may amount to tens of thousands of tonnes per year spread over the entire oceans.

There are two types of particles—iron-rich and silicate-rich—derived from different types of meteorite. Both form spherical particles called cosmic spherules, as the material is melted in the atmosphere and then solidifies in droplets.

Sediment Transport, Deposition and Accumulation

- Large particles sink more rapidly and need higher current speeds to resuspend them than smaller particles do.

- Large particles are not transported far by ocean currents but smaller particles can be carried long distances.

Orbital velocities in waves are much higher than ocean current speeds and waves resuspend sand-sized particles and move them long distances along the coast. However, when these particles are carried offshore to deeper water where wave orbit velocities are lower and wave motion does not extend to the seafloor, the particles are deposited and only the smallest particles are transported further by ocean currents.

Thus, large particles (which are primarily lithogenous) tend to collect in sediments of the continental shelf, whereas sediments accumulating far from land are generally very fine-grained and less likely to be dominated by lithogenous particles.

The smallest clay-sized particles form cohesive sediments that make the particles difficult to resuspend. Fine-grained cohesive muds often form in coastal areas such as wetlands that are protected from waves.

Turbidity Currents

Turbidity currents are swift-moving (70 km·hr −1 or more) slumps of sediment similar to avalanches. They can carry coarse grain-sized sediments to and across the deep ocean floor adjacent to the continental shelves.

Because large particles settle faster, turbidity currents leave layers of sediment called turbidite layers. In a turbidite layer, sediment grain size decreases upward toward the surface; finer-grained sediment layers appear above and below turbidite layers.

Accumulation Rates

Sediment accumulation rates are high near continents and much lower in the deep oceans far from land. All sediments are mixtures of particles from different origins. The composition of sediment at a given location is determined by the relative rates of accumulation of each type of material at that location.

Sediment accumulation rates in nearshore areas range from about 100 cm per 1000 years to extremes of several meters per year at some river mouths. On continental shelves and in marginal seas the rates are generally 10–100 cm per 1000 years. Rates in the deep oceans remote from land are much lower, about 0.1 cm per 1000 years.

Continental Margin Sediments

The continental shelves are covered in lithogenous sediments of larger grain sizes except in areas where currents are slow or production of biogenous particles is low.

On many continental shelves lithogenous sediment accumulates continuously as new material is supplied by coastal erosion and riverborne sediments.

On some continental shelves—such as that off the U.S. East Coast, where riverborne sediment is trapped in estuaries and lagoons and where currents speeds on the continental shelf are relatively high—little or no new sediment accumulates, especially on the outer part of the shelf, because inputs of lithogenous particles are low and particles of other types are generally small enough that they are transported away from these areas by currents.

In areas where no new sediment accumulates, the seafloor is covered by relict sediments. Relict sediments are sediments laid down when sea level was lower so that the deposition area was coastal and only covered by shallow water. Shells and other marine organism remains in relict sediment are often remains of species, such as some oysters that only live in very shallow coastal waters.

Distribution of Surface Sediments

The distribution of sediment types in surface sediments (currently accumulating material) reflects the proximity of lithogenous sediment sources, the productivity of the overlying waters and the type of organisms that are abundant, the seafloor depth, and the depth of the CCD.

Radiolarian Oozes

Radiolarian oozes accumulate in a region of high productivity that extends in a band across the deep oceans at the equator. However, radiolarian oozes do not accumulate in areas where the sedimentation rate of lithogenous material is much larger than that of radiolarian particles, as in most areas near the continents and in the Atlantic Ocean, where lithogenous sediment accumulation is higher than in other oceans.

Diatom Oozes

Diatoms are the dominant siliceous biogenous material except near the equator, where radiolaria dominate. Thus, diatom oozes are found in the deep oceans in areas of high productivity but only where inputs of lithogenous particles are low and where the seafloor is deeper than the CCD (so that calcareous biogenous particles are dissolved before they accumulate in the sediments).

Calcareous Sediments

In areas where the seafloor is shallower than the CCD and the inputs of lithogenous material are low, calcareous particles accumulate fast enough to be a major part of the sediments. These areas include oceanic plateaus, seamounts, and the flanks of the oceanic ridges.

Deep-sea Clays

In areas remote from land, deeper than the CCD, and where biological productivity in the overlying water is low, the only material that reaches the sediments in significant quantities is very fine-grained lithogenous particles transported large distances by currents and winds. These form slowly accumulating very fine-grained sediments called deep-sea clays, sometimes called red clays because the particles are reddish or brownish in color due to their iron oxide content.

Siliceous Red Clay Sediments

In the deep basins of the North and South Pacific, South Atlantic, and southern Indian Oceans there are transitional areas where sediments grade progressively between deep sea clays and diatom oozes.

Ice-rafted Sediments

Sediments carried to the oceans predominantly by glaciers accumulate in some areas of the Arctic Ocean, Bering Sea, and around Antarctica. They can contain pebbles and even larger particles, as icebergs originating from glaciers can carry these ice-rafted sediment particles far from land.

Terrigenous Sediments

Terrigenous sediments dominate in areas close to the mouths of rivers that carry large suspended sediment loads to the ocean, for example, in the northern Arabian Sea and the Bay of Bengal.

Hydrothermal Sediments

The central basin of the Red Sea is the only area where hydrothermal sediments are known to dominate. However, small areas of hydrothermal sediments also are found around hydrothermal vents.

The Sediment Historical Record

Sediments accumulate layer upon layer and preserve a history of changing deposition characteristics, although there is some mixing of sediments by bioturbation (churning of the upper layers of sediment by living organisms).

The sediment historical record can provide information about changes in depth of the seafloor, temperature of the overlying water, productivity of the overlying water, and the CCD, among other things. Reading the sediment historical record is difficult because so many different factors, including bioturbation, affect it and because the age of each layer must be determined precisely.

Sediment Age Dating

Ages of sediment layers are determined primarily by fossils and calibrated by radionuclide dating in sediments and rocks when possible. Some dating information can also be obtained from magnetic anomaly data and paleomagnetism.

Diagenesis

Physical and chemical changes called diagenesis occur in sediments over time as they are progressively buried.

The pore waters (water trapped between the mineral grains) are depleted in oxygen due to continued decomposition of organic matter. Eventually sulfides form, as the oxygen in sulfate is used by decomposers in place of the depleted oxygen. This allows metals that have soluble sulfides but insoluble hydrated oxides (e.g., iron, manganese) to dissolve. Silica and calcium carbonate are also dissolved progressively.

Oxygen diffuses slowly and is carried by bioturbation into sediments from the water above, whereas sulfides, dissolved silica, calcium, and carbonate ions formed from calcium carbonate dissolution can diffuse slowly upward within the sediment until they reach the oxygen diffusing down and are oxidized and precipitated. Pore waters are also squeezed out of the sediments as the sediments are compacted by the weight of the accumulated sediments above. Diagenesis is important because it can recycle nutrients from detritus in the sediments back to the water column.

Tectonic History in the Sediments

Because the type of seafloor sediment differs depending on depth, distance from the continents, and latitude, changes on a particular piece of crust can be used to reveal various aspects of the tectonic history.

For example, new crust near the oceanic ridges is shallower than the CCD, and sediments will contain calcium carbonate. However, as the crust moves away from the oceanic ridge, cools, and sinks isostatically, it descends below the CCD and newer sediment will not contain calcium carbonate.

Climate History in the Sediments

The past 170 million years of the Earth's climate history is preserved in sediments, primarily in biogenous particles. Different species grow in different temperature ranges. Thus, if we assume ancient species have similar temperature requirements to closely related species today, the remains of these species can tell us about the temperature range that existed at the place and time they were formed.

Past climate can also be determined from certain isotope ratios on biogenous particles. Oxygen consists primarily of two naturally occurring isotopes: O-16 and O-18. O-16 containing water evaporates slightly faster than O-18 containing water. When world climate is cold and there are more glaciers and ice sheets O-16 is transferred preferentially to this ice and the ratio of O-16 to O-18 in seawater goes down. Because the ratio of oxygen isotopes in marine organisms carbonate shell material is primarily determined by the ratio of the isotopes in the surrounding water, changes in the isotope ratio of calcareous sediments can be used to determine past climate temperatures.

Isotopes of an element also react at slightly different rates at different temperatures. The isotope ratios of most elements are relatively uniform throughout the oceans at any one point in history. The ratios of isotopes taken up by different species vary but, within a single species, the ratio depends only on the ratio in seawater and the temperature. As a result, isotope ratio differences between biogenous remains of a single species deposited at the same time but in different locations can reveal the temperature difference between the locations at the time the sediments were deposited.

Support for Extinction Theories in the Sediments

About 65 million years ago the last dinosaurs and more than half of all marine species became extinct. Other extinctions have occurred during the Earth's history. Evidence gathered from 65 million-year-old sediments supports a theory that this extinction was caused by a giant meteorite impact in the oceans near what is now the Yucatán Peninsula in Mexico.

Evidence of Impact at Chicxulub

The Yucatán Peninsula was about 500 m underwater 65 million years ago. Magnetic surveys show a 180 km diameter impact crater buried in the region now called Chicxulub. 65 million-year-old sediments in this and surrounding areas have a thick layer of unusual materials. Within this layer, the deepest sediments are rounded, very coarse-grained material called tektites, which are glassy material formed when rocks are melted by meteorite impacts. Above that, there is a layer of coarse-grained sediment that contains fossilized terrestrial plant matter and then progressively finer-grained material until, at the top of the anomalous layer, marine sediments normal for this region reappear.

This evidence can best be explained by a meteorite impact that caused a mega-tsunami that swept onto the continents and back, perhaps sloshing back and forth for days. This tsunami may have been caused by the impact itself or by an estimated magnitude 11 massive earthquake that the impact generated.

Evidence of other Impacts

There is evidence of a number of craters similar to the Chicxulub crater and associated anomalous sediment layers in various parts of the world that were possibly created by meteorite impacts at various times in the Earth's past. Some of these may also be associated with mass extinctions.

A crater of this nature found at the mouth of Chesapeake Bay was apparently created approximately 35 million years ago.

Ocean Basins

Ocean basin is any of several vast submarine regions that collectively cover nearly three-quarters of Earth's surface. Together they contain the overwhelming majority of all water on the planet and have an average depth of almost 4 km (about 2.5 miles). A number of major features of the basins depart from this average—for example, the mountainous ocean ridges, deep-sea trenches, and jagged, linear fracture zones. Other significant features of the ocean floor include aseismic ridges, abyssal hills, and seamounts and guyots. The basins also contain a variable amount of sedimentary fill that is thinnest on the ocean ridges and usually thickest near the continental margins.

General Features

While the ocean basins lie much lower than sea level, the continents stand high—about 1 km (0.6 mile) above sea level. The physical explanation for this condition is that the continental crust is light and thick while the oceanic crust is dense and thin. Both the continental and oceanic crusts lie over a more uniform layer called the mantle. As an analogy, one can think of a thick piece of styrofoam and a thin piece of wood floating in a tub of water. The styrofoam rises higher out of the water than the wood.

The ocean basins are transient features over geologic time, changing shape and depth while the process of plate tectonics occurs. The surface layer of Earth, the lithosphere, consists of a number of rigid plates that are in continual motion. The boundaries between the lithospheric plates form the principal relief features of the ocean basins: the crests of oceanic ridges are spreading centres where two plates move apart from each other at a rate of several centimetres per year. Molten rock material wells up from the underlying mantle into the gap between the diverging plates and solidifies into oceanic crust, thereby creating new ocean floor. At the deep-sea trenches, two plates converge, with one plate sliding down under the other into the mantle where it is melted. Thus, for each segment of new ocean floor created at the ridges, an equal amount of old oceanic crust is destroyed at the trenches, or so-called subduction zones. It is for this reason that the oldest segment of ocean floor, found in the far western Pacific, is apparently only about 200 million years old, even though the age of Earth is estimated to be at least 4.6 billion years.

The dominant factors that govern seafloor relief and topography are the thermal properties of the oceanic plates, tensional forces in the plates, volcanic activity, and sedimentation. In brief, the oceanic ridges rise about 2 km (1.2 miles) above the seafloor because the plates near these spreading centres are warm and thermally expanded. In contrast, plates in the subduction zones are generally cooler. Tensional forces resulting in plate divergence at the spreading centres also create block-faulted mountains and abyssal hills, which trend parallel to the oceanic ridges. Seamounts and guyots, as well as abyssal hills and most aseismic ridges, are produced by volcanism. Continuing sedimentation throughout the ocean basin serves to blanket and bury many of the faulted mountains and abyssal hills with time. Erosion plays a relatively minor role in shaping the face of the deep seafloor, in contrast to the continents. This is because deep ocean currents are generally slow (they flow at less than 50 cm [20 inches] per second) and lack sufficient power.

Morphology of Ocean Basins

Continental slopes typically have slope angles of between 28 and 108 and the continental rise is even less. Nevertheless, they are physiographically significant, as they contrast with the very low gradients of continental shelves and the flat ocean floor. Continental slopes extend from the shelf edge, about 200m below sea level, to the basin floor at 4000 or 5000 m depth and may be up to a hundred kilometres across in a downslope direction. Continental slopes are commonly cut by submarine canyons, which, like their counterparts on land, are steep-sided erosional features. Submarine canyons are deeply incised, sometimes into the bedrock of the shelf, and may stretch all the way back from the shelf edge to the shoreline. They act as conduits for the transfer of water and sediment from the shelf, sometimes feeding material directly from a river mouth. The presence of canyons controls the formation and position of submarine fans. The generally flat surface of the ocean floor is interrupted in places by seamounts, underwater volcanoes located over isolated hotspots. Seamounts may be wholly submarine or may build up above water as volcanic islands, such as the Hawaiian island chain in the central Pacific. As subaerial volcanoes they can be important sources of volcaniclastic sediment to ocean basins. The flanks of the volcanoes are commonly unstable and give rise to very largescale submarine slides and slumps that can involve several cubic kilometres of material. Bathymetric mapping and sonar images of the ocean floor around volcanic islands such as Hawaii in the Pacific and the Canary Islands in the Atlantic have revealed the existence of very large-scale slump features. Mass movements on this scale would generate tsunami around the edges of the ocean, inundating coastal areas. The deepest parts of the oceans are the trenches formed in regions where subduction of an oceanic plate is occurring. Trenches can be up to 10,000 m deep. Where they occur adjacent to continental margins (e.g. the Peru–Chile Trench west of South America) they are filled with sediment supplied from the continent, but mid-ocean trenches, such as the Mariana Trench in the west Pacific, are far from any source of material and are unfilled, starved of sediment.

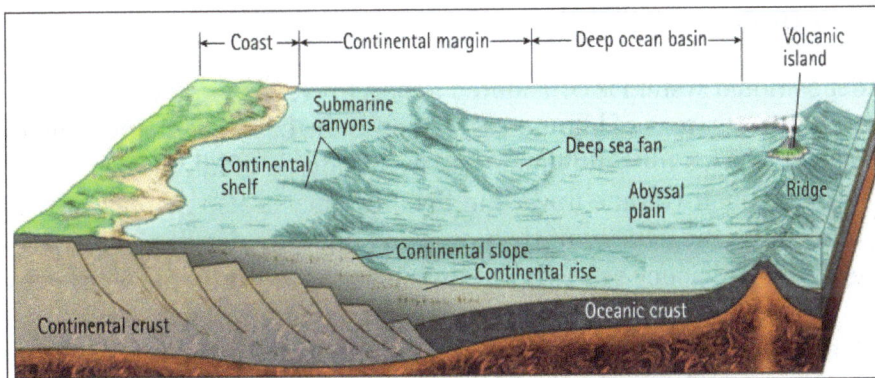

Depositional Processes in Deep Seas

Deposition of most clastic material in the deep seas is by mass-flow processes. The most common are debris flows and turbidity currents, and these form part of a spectrum within which there can be flows with intermediate characteristics.

Debris-flow Deposits

Remobilisation of a mass of poorly sorted, sediment rich mixture from the edge of the shelf or the top of the slope results in a debris flow, which travels down the slope and out onto the basin plain. Unlike a debris flow on land an underwater flow has the opportunity to mix with water and in doing so it becomes more dilute and this can lead to a change in the flow mechanism and a transition to a turbidity current. The top surface of a submarine debris flow deposit will typically grade up into finer deposits due to dilution of the upper part of the flow. Large debris flows of material are known from the Atlantic off northwest Africa and examples of thick, extensive debris-flow deposits are also known from the stratigraphic record. Debris-flow deposits tens of metres thick and extending for tens of kilometres are often referred to as megabeds.

Turbidites

Dilute mixtures of sediment and water moving as mass flows under gravity are the most important mechanism for moving coarse clastic material in deep marine environments. These turbidity currents carry variable amounts of mud, sand and gravel tens, hundreds and even over a thousand kilometres out onto the basin plain. The turbidites deposited can range in thickness from a few millimetres to tens of metres and are carried by flows with sediment concentrations of a few parts per thousand to 10%. Denser mixtures result in high-density turbidites that have different characteristics to the 'Bouma Sequences' seen in low- and medium-density turbidites. Direct observation of turbidity currents on the ocean floor is very difficult but their effects have been monitored on a small number of occasions. In November 1929 an earthquake in the Grand Banks area off the coast of Newfoundland initiated a turbidity current. The passage of the current was recorded by the severing of telegraph cables on the sea floor, which were cut at different times as the flow advanced. Interpretation of the data indicates that the turbidity current travelled at speeds of between 60 and 100 km. Also, the deposits of recent turbidity flows have been mapped out, for example, in the east Atlantic off the Canary Islands a single turbidite deposit has been shown to have a volume of 125 km cube.

High and Low-efficiency Systems

A deep marine depositional system is considered to be a low-efficiency system if sandy sediment is carried only short distances (tens of kilometres) out onto the basin plain and a high-efficiency system if the transport distances for sandy material are hundreds of kilometres. High-volume flows are more efficient than small-volume flows and the efficiency is also increased by the presence of fines that tend to increase the density of the flow and hence the density contrast with the seawater. The deposits of low-efficiency systems are therefore concentrated near the edge of the basin, whereas muddier, more efficient flows carry sediment out on to the basin plain. The high-efficiency systems will tend to have an area near the basin margin called a bypass zone where sediment is not deposited, and there may be scouring of the underlying surface, with all the deposition concentrated further out in the basin.

Initiation of Mass Flows

Turbidity currents and mass flows require some form of trigger to start the mixture of sediment and water moving under gravity. This may be provided by an earthquake as the shaking generated by a seismic shock can temporarily liquefy sediment and cause it to move. The impact of large storm waves on shelf sediments may also act as a trigger. Accumulation of sediment on the edge of the shelf may reach the point where it becomes unstable, for example where a delta front approaches the edge of a continental shelf. High river discharge that results in increased sediment supply can result in prolonged turbidity current flow as sediment-laden water from the river mouth flows as a hyperpycnal flow across the shelf and down onto the basin plain. Such quasi-steady flows may last for much longer periods than the instantaneous triggers that result in flows lasting just a few hours. A fall in sea level exposes shelf sediments to erosion, more storm effects and sediment in-stability that result in increased frequency of turbidity currents.

Composition of Deep Marine Deposits

The detrital material in deep-water deposits is highly variable and directly reflects the sediment source area. Sand, mud and gravel from a terrigenous source are most common, occurring offshore continental margins that have a high supply from fluvial sources. Material that has had a short residence time on the shelf will be similar to the composition of the river but extensive reworking by wave and tide processes can modify both the texture and the composition of the sediment before it is redeposited as a turbidite. Sandstone deposited by a turbidity current can therefore be anything from a very immature, lithic wacke to a very mature quartz arenite. Turbidites composed wholly or partly of volcaniclastic material occur in seas offshore of volcanic provinces. The deep seas near to carbonate shelves may receive large amounts of reworked shallow-marine carbonate sediment, redeposited by turbidity currents and debris flows into deeper water: recognition of the redeposition process is particularly important in these cases because the sediment will contain bioclastic material that is characteristic of shallow water environments. Because there is this broad spectrum of sandstone compositions in deep-water sediments, the use of the term 'greywacke' to describe the character of a deposit is best avoided: it has been used historically as a description of lithic wackes that were deposited as turbidites and the distinction between composition and process became confused as the terms turbidite and greywacke came to be used almost as synonyms. 'Greywacke' is not part of the Pettijohn classification of sandstones and it no longer has any widely accepted meaning in sedimentology.

Exploration of the Ocean Basins

Mapping the characteristics of the ocean basin has been difficult for several reasons. First, the oceans are not easy to travel over; second, until recent times navigation has been extremely crude, so that individual observations have been only loosely correlated with one another; and, finally, the oceans are opaque to light—that is, the deep seafloor cannot be seen from the ocean surface. Modern technology has given rise to customized research vessels, satellite and electronic navigation, and sophisticated acoustic instruments that mitigate some of these problems.

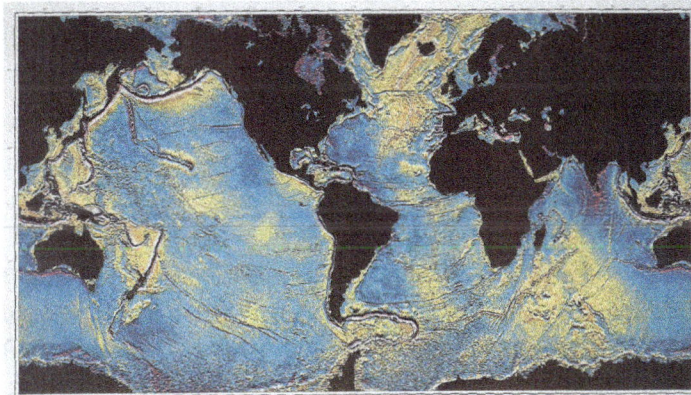

The Challenger Expedition, mounted by the British in 1872–76, provided the first systematic view of a few of the major features of the seafloor. Scientists aboard the HMS Challenger determined ocean depths by means of wire-line soundings and discovered the Mid-Atlantic Ridge. Dredges brought up samples of rocks and sediments off the seafloor. The main advance in mapping, however, did not occur until sonar was developed in the early 20th century. This system for detecting the presence of objects underwater by acoustic echo provided marine researchers with a highly useful tool, since sound can be detected over several thousands of km in the ocean (visible light, by comparison, can penetrate only 100 metres [about 330 feet] or so of water).

Gravity map of Earth's ocean surface, computed from radar-altimetry measurements made from orbit by the U.S. satellite Seasat in 1978. Because the ocean surface is deformed by the varying gravitational attraction of the underlying marine topography, such maps sensitively mirror seafloor features and have been valuable in identifying previously uncharted seamounts, ridges, and fracture zones.

Modern sonar systems include the Seabeam multibeam echo sounder and the GLORIA scanning sonar. They operate on the principle that the depth (or distance) of the seafloor can be determined by multiplying one-half the elapsed time between a downgoing acoustic pulse and its echo by the speed of sound in seawater (about 1,500 metres [4,900 feet] per second). Such multifrequency sonar systems permit the use of different pulse frequencies to meet different scientific objectives. Acoustic pulses of 12 kilohertz (kHz), for example, are normally employed to measure ocean depth, while lower frequencies—3.5 kHz to less than 100 hertz (Hz)—are used to map the thickness of sediments in the ocean basins. Very high frequencies of 100 kHz or more are employed in side-scanning sonar to measure the texture of the seafloor. The acoustic pulses are normally generated by piezoelectric transducers. For determining subbottom structure, low-frequency acoustic pulses are produced by explosives, compressed air, or water-jet implosion. Near-bottom sonar systems, such as the Deep Tow of the Scripps Institution of Oceanography, produce even more detailed images of the seafloor and subbottom structure. The Deep Tow package contains both echo sounders and side-scanning sonars, along with associated geophysical instruments, and is towed behind a ship at slow speed 10 to 100 metres (33 to 330 feet) above the seafloor. It yields very precise measurements of even finer-scale features than are resolvable with Seabeam and other comparable systems.

Another notable instrument system is ANGUS, a deep-towed camera sled that can take thousands of high-resolution photographs of the seafloor during a single day. It has been successfully used in the detection of hydrothermal vents at spreading centres. Overlapping photographic images make it possible to construct photomosaic strips about 10 to 20 metres (33 to 66 feet) wide that reveal details on the order of centimetres.

Three major navigation systems are in use in modern marine geology. These include electromagnetic systems such as loran and Earth-orbiting satellites. Acoustic transponder arrays of two or more stations placed on the seafloor a few kilometres apart are used to navigate deeply towed instruments, submersibles, and occasionally surface research vessels when detailed mapping is conducted in small areas. These systems measure the distance between the instrument package and the transponder sites and, using simple geometry, compute fixes accurate to a few metres. Although the individual transponders can be used to determine positions relative to the array with great accuracy, the preciseness of the position of the array itself depends on which system is employed to locate it.

Earth-orbiting satellites such as Seasat and Geosat have uncovered some significant topographic features of the ocean basins. Seasat, launched in 1978, carried a radar altimeter into orbit. This device was used to measure the distance between the satellite path and the surfaces of the ocean and continents to 0.1 metre (0.3 foot). The measurements revealed that the shape of the ocean surface is warped by seafloor features: massive seamounts cause the surface to bulge over them because of gravitational attraction. Similarly, the ocean surface downwarps occur over deep-sea trenches. Using these satellite measurements of the ocean surface, William F. Haxby computed the gravity field there. The resulting gravity map provides comprehensive coverage of the ocean surface on a 5'-by-5' grid that depicts five nautical miles on each side at the Equator). Coverage as complete as this is not available from echo soundings made from ships. Because the gravity field at the ocean surface is a highly sensitive indicator of marine topography, this map reveals various previously uncharted features, including seamounts, ridges, and fracture zones, while improving the detail on other known features. In addition, the gravity map shows a linear pattern of gravity anomalies that cut obliquely across the grain of the topography. These anomalies are most pronounced in the Pacific basin; they are apparently about 100 km (about 60 miles) across and some 1,000 km (about 600 miles) long. They have an amplitude of approximately 10 milligals (0.001 percent of the Earth's gravity attraction) and are aligned west-northwest—very close to the direction in which the Pacific Plate moves over the mantle below.

Deep-sea Sediments

The ocean basin floor is everywhere covered by sediments of different types and origins. The only exception are the crests of the spreading centres where new ocean floor has not existed long enough to accumulate a sediment cover. Sediment thickness in the oceans averages about 450 metres (1,500 feet). The sediment cover in the Pacific basin ranges from 300 to 600 metres (about 1,000 to 2,000 feet) thick, and that in the Atlantic is about 1,000 metres (3,300 feet). Generally, the thickness of sediment on the oceanic crust increases with the age of the crust. Oceanic crust adjacent to the continents can be deeply buried by several kilometres of sediment. Deep-sea sediments can reveal much about the last 200 million years of Earth history, including seafloor spreading, the history of ocean life, the behaviour of Earth's magnetic field, and the changes in the ocean currents and climate.

The study of ocean sediments has been accomplished by several means. Bottom samplers, such as dredges and cores up to 30 metres (about 100 feet) long, have been lowered from ships by wire to retrieve samples of the upper sediment layers. Deep-sea drilling has retrieved core samples from the entire sediment layer in several hundred locations in the ocean basins. The seismic reflection method has been used to map the thickness of sediments in many parts of the oceans. Besides thickness, seismic reflection data can often reveal sediment type and the processes of sedimentation.

Sediment Types

Deep-sea sediments can be classified as terrigenous, originating from land; as biogenic, consisting largely of the skeletal debris of microorganisms; or as authigenic, formed in place on the seafloor. Pelagic sediments, either terrigenous or biogenic, are those that are deposited very slowly in the open ocean either by settling through the volume of oceanic water or by precipitation. The sinking rates of pelagic sediment grains are extremely slow because they ordinarily are no larger than several micrometres. However, fine particles are normally bundled into fecal pellets by zooplankton, which allows sinking at a rate of 40 to 400 metres (130 to 1,300 feet) per day.

Sedimentation Patterns

The patterns of sedimentation in the ocean basins have not been static over geologic time. The existing basins, no more than 200 million years old, contain a highly variable sedimentary record. The major factor behind the variations is plate movements and related changes in climate and ocean water circulation. Since about 200 million years ago, a single vast ocean basin has given way to five or six smaller ones. The Pacific Ocean basin has shrunk, while the North and South Atlantic basins have been created. The climate has changed from warm and mild to cool, stormy, and glacial. Plate movements have altered the course of surface and deep ocean currents and changed the patterns of upwelling, productivity, and biogenic sedimentation. Seaways have opened and closed. The Strait of Gibraltar, for example, was closed off about 6 million years ago, allowing the entire Mediterranean Sea to evaporate and leave thick salt deposits on its floor. Changes in seafloor spreading rates and glaciations have caused sea levels to rise and fall, greatly altering the deep-sea sedimentation pattern of both terrigenous and biogenic sediments. The calcite compensation depth (CCD), or the depth at which the rate of carbonate accumulation equals the rate of carbonate dissolution, has fluctuated more than 2,000 metres (about 6,600 feet) in response to changes in carbonate supply and the corrosive nature of ocean bottom waters. Bottom currents have changed, becoming erosive or nondepositional in some regions to produce geological unconformities (that is, gaps in the geological record) and redistribute enormous volumes of sediment to other locations. The Pacific Plate has been steadily moving northward, so that biogenic sediments of the equatorial regions are found in core samples taken in the barren North Pacific.

References

- Ponta, f.l.; p.m. jacovkis (april 2008). "marine-current power generation by diffuser-augmented floating hydro-turbines". Renewable energy. 33 (4): 665–673. Doi:10.1016/j.renene.2007.04.008. Retrieved 2011-04-12.

- Ocean-current, science: britannica.com, Retrieved 3 August, 2019

- Hammons, thomas (2011). Electricity infrastructures in the global marketplace. Bod – books on demand. Isbn 978-9533071558.

- Sea+waves: encyclopedia2.thefreedictionary.com, Retrieved 19 May, 2019

- Battisti, david s., 2000: "developing a theory for enso," ncar advanced study program, "archived copy". Archived from the original on 2010-06-10. Retrieved 2010-08-21.cs1 maint: archived copy as title (link)

- Ocean-floor-sediments, oc-po: waterencyclopedia.com, Retrieved 28 April, 2019

- Ocean-basin, science: britannica.com, Retrieved 8 August , 2019

Permissions

Index